A Naturalist's Guide to Wetland Plants

A Naturalist's Guide to
Wetland Plants

An *Ecology for* Eastern North America

Donald D. Cox
Illustrations by Shirley A. Peron

Syracuse University Press

The paper used in this publication meets the minimum requirements of
American National Standard for Information Sciences—Permanence of
Paper for Printed Library Materials, ANSI Z39.48–1984∞™

Library of Congress Cataloging-in-Publication Data

Cox, Donald D.

A naturalist's guide to wetland plants : an ecology for eastern North
America / Donald D. Cox ; illustrations by Shirley A. Peron.— 1st ed.

p. cm.

Includes bibliographical references (p.).

ISBN 0–8156–0740–7 (pbk.)

1. Wetland plants—East (U.S.)—Identification. 2. Wetland
plants—North America—Identification. 3. Wetland plants—Ecology—East
(U.S.) 4. Wetland plants—Ecology—North America. I. Title.

QK115 .C72 2002

581.7′68′0974—dc21

2002004522

Manufactured in the United States of America

To Barbara

With love and gratitude
for editorial patience and creativity

Donald D. Cox studied at Marshall University, received a Ph.D. from Syracuse University, and was a professor of biology for forty-one years. His publications include *Some Postglacial Forests in Central and Western New York State*, *The Context of Biological Education: The Case for Change*, *Common Flowering Plants of the Northeast*, and *Seaway Trail Wildguide to Natural History*.

Contents

Illustrations

Acknowledgments

I am indebted to Dr. Kenneth Heilman for reading the section on poisonous plants. I wish to thank Barbara Cox and Shirley Peron for providing many helpful suggestions. To Brian McDonald I extend thanks for information on Cranberry Glades. I especially wish to thank Sharon Doerr for carefully reading the manuscript and offering suggestions for improvement. I am grateful for the courtesy and cooperation from the staff of Penfield Library at SUNY-Oswego. I wish to express my appreciation to the staff at Syracuse University Press for creative suggestions in the early phases, and for professional guidance throughout the publication process.

For the botanical and common names of plants I have, when applicable, used those presented in the second edition (1991) of the *Vascular Plants of Northeastern United States and Adjacent Canada* by Henry A. Gleason and Arthur Cronquist.

Introduction

Wetland habitats are the most endangered ecosystems in the world. Although in the past half century the attitude toward them has improved, wetlands are still shrinking. This book is about the plants that inhabit freshwater wetlands. Each species has unique features that enable it to survive in its environment. Each is a vital part of its ecosystem, and wetland ecosystems are essential to the ecology of the earth.

In the following chapters, for ease of reading, technical terminology has been kept to a minimum. Some terms used to describe plants, although not highly technical, may have special meanings that are unfamiliar to the reader. For these, a glossary has been included. As interest in and knowledge of plants grows, it becomes clear that relying on common names can be confusing because every region may have a different name for the same plant. For this reason, the botanical name is given in parentheses. With a little practice these will become as easy to use as common names, and they are much more reliable. The botanical name for a species is the same all over the world.

The way wetlands function as ecosystems and their role in the ecology of the earth are discussed in chapter 1. This chapter includes descriptions of several important wetlands in eastern North America. In chapter 2, brief descriptions are given for each of the major types of plant groups that grow in freshwater wetlands. The remainder of the book focuses mainly on one group, the flowering plants or angiosperms. Emphasis has been placed on field observations, the intention being to provide descriptions and drawings that will aid in identification. With a little effort, an alert observer can learn to recognize the specific characteristics of many species. In a given region of North America, there may be several thousand species of plants. Trying to learn all of these is a daunting challenge, and the beginner may wish to begin by learning the plants in a limited group. For example, he or she could start by learning the trees, shrubs, toxic, medicinal, or edible plants in a wetland. All of these groups and others are described in the following chapters.

Wetland plants have special characteristics that enable them to survive in this habitat. These characteristics are described in chapter 3. In chapters 4 and 5, swamp and bog ecology and the main plants that inhabit them are described. Chapter 6 presents toxic, medicinal, and edible plants of freshwater wetlands. Chapter 7 describes the changes that take place through the seasons. It includes a section on aquatic plants that may become weed pests. Chapter 8 details inexpensive methods for those interested in collecting and preserving plants. Chapter 9 is offered for those who wish to go a step beyond identifying or collecting plants. It includes activities, projects, and investigations.

Observing and learning about wetlands can be entertaining and educational. Since plants cannot run away, they can be observed or studied in detail in their natural habitats. While this is convenient for all who study or observe them, it makes plants vulnerable to all sorts of destructive forces. Their greatest threat is disruption of habitats resulting from human activities. The destruction of natural areas is increasing as the human population grows, making habitat conservation an urgent priority.

The ability to identify plants has its own reward in personal satisfaction. Recognizing plant species when visiting a strange wetland is like seeing old friends. It is comforting to know that even in a strange environment there are familiar "faces." In addition, as one travels through the countryside, being aware of the contribution wetland communities make to the landscape gives it a richer meaning and enhances the enjoyment of viewing it. It is the aim of this book to give the reader a broader understanding and a greater appreciation for the plants of wetland ecosystems.

Three additional publications are available as parts of the "Naturalist's Guides" series. These are *A Naturalist's Guide to Forest Plants*, *A Naturalist's Guide to Meadow Plants*, and *A Naturalist's Guide to Seashore Plants*. Topics covered in these publications include plant lore, ecology, and tips for plant identification; poisonous, hallucinogenic, medicinal, and wild food plants;

and collecting and preserving plants. In addition, there are activities, projects, plant investigations, and thought stimulators. For naturalists and other lovers of the out-of-doors, these books not only provide ecological backgrounds but can also serve as supplementary field guides.

A Naturalist's Guide to Wetland Plants

1

Wetlands as Ecosystems

What Are Wetlands?

Swamp, bog, marsh, fen, muskeg, wet meadow, mire, moor, bottomland, slough, heath: the number of names for wet places reflects the difficulty in constructing a definition of wetlands that fits all situations. In addition to encompassing different climates, locations, and plants, wetlands may be transition zones between deep water and upland areas, and it is sometimes hard to determine the exact location of the margins. Most technical definitions include information about three components: water, soil, and vegetation. In nontechnical terms, a wetland is an area where the soil is saturated with or covered by water and supports hydrophytic plants—plants that can grow in water or very wet soil—for a good portion of the year.

Before the 1960s wetlands were usually viewed as wastelands, sources of disease, unfit for farming or any other useful purpose. From early in the nineteenth century through the mid-twentieth century, the destruction of wetlands was considered an acceptable and proper practice. Consequently by the mid 1950s, more than half of the wetlands present in the 1600s in the United States had been destroyed. They had been ditched, drained, and filled to make way for agriculture, housing projects, shopping malls, and other construction. In the twenty-year period from 1955 to 1975, there was a decrease of about 10 percent in wetland areas in the United States. Beginning in the mid 1970s, state and federal protective legislation slowed, but did not stop, the destruction of this valuable resource. Since then, by current estimates, wetland areas have continued to shrink by about 5 percent per year.

Technically, five types of wetlands are recognized by ecologists: marine, estuarine, lacustrine, riverine, and palustrine. Marine and estuarine wetlands are coastal and are associated with salt water. These are discussed in another publication in this series. Lacustrine wetlands are lake related; riverine wetlands occur along rivers and streams; and palustrine wetlands refer to what are commonly called marshes, swamps, and bogs. The latter three types of freshwater wetlands make up at least 90 percent of the wetlands in the United States and Canada.

Palustrine wetlands are the main focus of this book, but the lines of distinction between these three ecosystems are blurred because the same plant species may occur in all three. Thus, to simplify terminology, two types of wetlands will be recognized: swamps and bogs. The basic physical difference between the two is in the nature of the underlying base or substrate. Beneath a relatively thin layer of organic debris, swamps have mineral soil as a base; bogs have a substrate of peat that may be a few to many feet (dm) thick. Swamps usually have a stream flowing through them or are drained by a stream and are flushed annually by high water. Bogs form mostly in depressions that have no outlets and are thus not drained. Both swamps and bogs are inhabited by populations of plants and animals that can survive in no other habitat.

What Good Are They?

Ecologists recognized the value of wetlands long before protective legislation was enacted by federal and state governments. However, it was the combined activism of hunters, wildlife watchers, nature photographers, wetland researchers, and lovers of the natural world who raised the awareness in legislative bodies of the need for protection. After the Clean Water Act was passed by Congress in 1977, several states enacted laws protecting wetlands, and at this writing others are considering the problem. In spite of legal protection, the acreage and quality of this resource is continuing to decline, by one estimate, at the rate of 386 to 772 square miles (1,000–2,000 sq. km) per year. Another method of calculating the loss has been presented by Janet Abramovitz of the Worldwatch Institute. During the years between the 1780s and 1980s, wetlands were destroyed at the average rate of 60 acres (24 ha) per hour for each hour of this two-hundred-year period.

Most of the remaining wetlands, 74 percent of which are privately owned, are considered by the U.S. Fish and Wildlife Service to be prime wildlife habitats.

Habitat for Living Things

Many species of plants and animals can survive only in wetlands. One-third of all North American bird species rely directly on wetlands for some resource. These areas provide unique environmental conditions that are required by an estimated nine hundred species of wildlife at some stage in their lives. As the result of habitat destruction, a disproportionately high percentage of these species are threatened or endangered. By one estimate, 26 percent of the plants and 45 percent of the animals on threatened or endangered lists are directly or indirectly dependent on wetlands. These include 15 percent of endangered mammals, 31 percent of threatened or endangered bird species, and 31 percent of threatened or endangered reptiles in the United States. If wetlands continue to shrink, it is likely that, in the future, many of these species will become extinct.

There are those who promulgate the idea that extinctions have been occurring since life began and current extinctions are normal events of nature. To be sure, extinctions have been common throughout geologic time, but the current number far exceeds that which would be expected to occur naturally. Even more important than the actual *number* of species that are threatened with extinction is the *rate* at which they are becoming extinct. The current rate is one hundred to one thousand times greater than is natural. If the species that are now threatened are eliminated in the next century, the rate will be one thousand to ten thousand times greater than the natural rate of extinction.

Flood Protection

One of the great values of wetlands is the role they play in flood prevention. They are depressions that serve as catch basins in times of heavy precipitation. Upstream wetlands catch and hold the runoff, thus reducing the level of the peak flood stage and the accompanying flood damage. Investigators have found that the greater the acreage of wetlands along the course of a stream, the less severe and the shorter the duration of flooding in times of excessive precipitation. A study by the U.S. Army Corps of Engineers in the St. Charles River Valley in Massachusetts concluded that drainage of a 13-square-mile (34 sq. km) forested wetland would increase flood damage on downstream areas by $17 million per year.

Groundwater Reservoir

If a hole is drilled into the ground almost anywhere in North America, a zone will eventually be reached that is saturated with water. This is the ground water zone, and its top surface is the water table. Many rural homes, towns, and cities sink wells into the groundwater for human use. The source of water in this underground reservoir is melting snow and rainwater that has percolated down from the surface. During extended periods of low precipitation, the groundwater level drops and wells may go dry. In these circumstances, percolating water from wetlands may reduce the extent of drop in groundwater levels. They thus provide an insurance policy against long-term droughts.

Recreation, Education, and Research

Wetlands offer great opportunities for many types of recreation. Hunting (mainly waterfowl), trapping, and fishing are well-known activities, as are nature photography, bird watching, and canoeing. Nature lovers are attracted to areas in which the influence of humans has been minimal and the environment can be observed in its natural state. Guided boat tours are among the most popular attractions in wetlands such as Everglades National Park, Okefenokee Swamp, and the Great Dismal Swamp.

Wetlands are also excellent areas to observe and demonstrate many of the scientific principles of environmental education; the importance of wetland research cannot be overstated. Unique wetland plants such as carnivorous plants and rare native orchids can be observed. The principles of ecological succession, energy relationships between plants and animals, and nutrient recycling are easier to demonstrate in wetlands than in other environments. Our knowledge of these ecosystems is incomplete but what has been learned to date has contributed to our understanding of the earth's ecosystems.

Water Purification and Waste Disposal

It has been demonstrated that plants can effectively purify water. Because of their cleansing action on polluted water, wetlands have been called the "kidneys of the landscape." Two essential components of waste treatment and water purification are microorganisms and oxygen. Plants produce an abundance of oxygen during photosynthesis. Hydrophytes, or wetland plants, have aeration systems that conduct oxygen to all underwater parts, even to those anchored in oxygen-free sediments. Oxygen in the air spaces of underwater roots, stems, and leaves leaks from the surfaces of these organs into the surrounding spaces. These surfaces provide a large aerated area where microorganisms can function to decompose organic wastes. This is the same process that takes place in sewage treatment plants, but it occurs over a much larger surface area. For some communities, it may be more cost effective to maintain a wetland than to build and operate an expensive sewage treatment plant. It has been estimated that a one-hundred-acre (250 ha) marsh-pond wetland could effectively handle the domestic sewage from a community of ten thousand people.

Wetlands and the Biosphere

Human Economics

Scientific investigations have revealed that wetlands are much more important to human welfare than was formerly believed. They are not the useless wastelands that unfortunately they are still thought to be by many people. In addition to the values described above, wetlands have great commercial significance. As much as 95 percent of the commercially harvested fish and shellfish spend part of their life cycles in wetlands. These organisms are the base of a business enterprise that generates more than $5 billion per year and employs at least 330,000 people in the United States. It has been estimated by G. Tyler Miller Jr. (1992) that an acre (0.4 ha) of coastal wetlands is worth $83,000 if its food production and recreational value are included. Compare this with an acre (0.4 ha) of prime corn or wheat farmland in Kansas with a top value of $1,200 and annual crop production value of $600.

Many communities in Florida obtain their water supplies by sinking wells into water holding underground strata called aquifers. In one instance, a 557,500 acre (223,000 ha.) swamp was calculated to have a value of $25 million annually for its service of storing water and recharging the aquifer. In the relatively dry midwestern United States and Canada, an acre (0.4 ha) of prairie pothole marshes can store as much as one and a half million gallons of water. The prairie pothole marshes will be discussed in a later section of this chapter.

Productivity

Biological productivity is the total amount of living tissue or biomass produced by an ecosystem in a specific amount of time. Wetlands are among the most productive ecosystems on earth, equaled only by the tropical rain forests. The high density of vegetation in wetlands can convert large amounts of radiant energy from the sun to chemical energy and store it as carbohydrates and other organic compounds. In addition, when wetland plants die, they are efficiently decomposed and recycled by the rich bacterial and fungal populations of the substrate. Also contributing to the fertility in most wetlands is periodic flooding, which brings a new layer of nutrient-rich sediments. A reflection of this fertility can be seen in a cattail marsh, which is one of the most productive wetlands. Cattail marshes may have yields as high as twelve tons of biomass per acre per year.

A contributing factor to the high productivity of wetlands is that they support many species whose metabolism categorizes them as C_4 plants. These have a different internal leaf structure and use carbon dioxide more efficiently than other plants. They are called C_4 plants because the first product of photosynthesis is a compound with four carbon atoms while in most species (C_3 plants) the first product of photosynthesis is a three-carbon compound. Species with C_4 photosynthesis have greater rates of photosynthesis per leaf surface area per hour, greater daily growth rates, and greater daily productions of biomass. There are other physiological differences between these two types of species, and it should be noted that all freshwater wetland plants do not have C_4 photosynthesis and some C_4 plants grow in places other than wetlands.

Approximately 6 percent of the earth's land surface can be classified as wetlands. Although this is a small percentage, wetlands exert a disproportionately large influence on the stability of the biosphere, that portion of the earth that sustains life. The nature of the substrate and the high rates of photosynthesis and recycling make wetlands vital components in the functioning of the total earth ecosystem. All life on the planet is regulated by the concentrations of key substances in the biosphere. Atmospheric components—including nitrogen, oxygen, carbon dioxide, and water—and other essential substances are constantly recycled keeping the biosphere suitable for living things.

Nitrogen

In the nitrogen cycle, gaseous nitrogen in the atmosphere is converted to forms that can be used by plants, then through natural pathways is eventually returned to the atmosphere. Nitrogen is one of the building blocks of proteins, including genetic proteins DNA and RNA, and is essential for all living things. Green plants cannot use the gaseous form that makes up about 78 percent of the atmosphere; instead they absorb nitrogen from the soil in the form of nitrogen compounds called nitrates. There are at least two natural sources of nitrates in the soil. Microorganisms called nitrogen-fixing bacteria convert gaseous nitrogen to nitrates. The first plants on land to evolve from the sea required nitrates for survival, and it was probably provided by these bacteria. The other major source of nitrates in the soil results from the decomposition of the dead bodies of plants and animals. Animals must also have nitrogen to build proteins, and they get it by eating plants and other animals.

Soils that are used for the cultivation of crops are often poor in nitrates, so farmers use fertilizers to assure better harvests. Since the compounds in these fertilizers are soluble, the rainwater that percolates into the groundwater and runs off the fields may have relatively high concentrations of nitrates. These may become pollutants in human drinking water. This is where wetlands enter the picture. They are home to vast numbers of denitrifying bacteria that are capable of breaking down nitrates and releasing gaseous nitrogen back into the atmosphere. The bacteria are anaerobic (they do not require oxygen to live) and are well suited for this often oxygen-free habitat. In addition to playing an important role in maintaining the stability of the nitrogen cycle, wetlands also provide a service to humanity by reducing potentially hazardous concentrations of nitrates in water that might become groundwater used for human consumption.

Oxygen

One environmental interaction shared by all humans and probably the one most often taken for granted is breathing the air. Yet if there is one essential chemical that could be called the most important, it is the oxygen in the air. Humans can survive for approximately a month without food and about a week without water but only a few minutes without oxygen.

The early atmosphere of the earth contained almost no oxygen. Billions of years of photosynthesis have resulted in an atmospheric oxygen content of about 21 percent. The concentration has probably been at this level throughout the evolution of mammals. It may be that a drop of only a few percent would threaten the survival of these organisms, including humans.

Organisms that require oxygen use it to convert the chemical energy of food into energy necessary to live, grow, and move. Although they do not usually move the way animals do, most plants grow throughout their lives and require energy to produce leaves, flowers, fruits, and seeds. The process of converting food into energy is called cellular respiration, and it takes place in all living things. Carbon dioxide is a by-product of respiration. During daylight hours, plants carry on photosynthesis, generating more oxygen than they use in respiration. The surplus diffuses into the atmosphere. In addition to that released in respiration, they must absorb additional carbon dioxide from the atmosphere. In the dark, plants do not generate oxygen, and like most other living things they release carbon dioxide into their environment and absorb oxygen from it.

The only important source of oxygen on earth is from the process of photosynthesis. It is photosynthesis that keeps oxygen at its present level in the atmosphere. The rate of oxygen production in wetlands is unsurpassed by any other vegetation type on earth and is equaled only by the tropical rain forests. The quantity of oxygen added to the atmosphere by wetland plants and all other photosynthetic organisms completely replenishes the oxygen in the biosphere about every two years.

Carbon Dioxide

Wetlands have a stabilizing influence on the carbon cycle. Through very high rates of photosynthesis, they remove great quantities of carbon dioxide from the atmosphere, then, except in peat bogs, return it by respiration and decay. In the biosphere, these processes balance each other and have kept carbon dioxide at about the same concentration for thousands of years. It currently makes up about .03 percent of the atmosphere, but it has not always been at this level. During the time of the Coal Age, about 300 million years ago, there may have been several thousand times more carbon dioxide in the air than today. The level was reduced to the present concentration by photosynthesis in the Coal Age swamp forests that were converted into great deposits of coal, oil, and natural gas. These are the fossil fuels that provide 90 percent of the energy used by humans today.

The atmospheric carbon dioxide/oxygen balance maintained by photosynthesis and respiration/decay began to change with the invention of the steam engine and the initiation of the industrial revolution about 1800. Steam-operated machines using coal for fuel began to add more carbon dioxide to the air than was being removed by photosynthesis. At first the accumulation was very slight, but it increased as the use of coal, oil, and natural gas escalated. The invention and use of the internal combustion engine resulted in a great increase in the emission of carbon dioxide. In the past one hundred years, the concentration of carbon dioxide in the air has increased by 25 percent. The greatest increase has been in the last fifty years. It is now at a level estimated to be the highest in 130,000 years.

Carbon dioxide is added to the atmosphere by a number of modern activities. Electricity is the most common form of energy used in much of the world. In the United States and many other countries, 90 percent of the electricity is produced by burning fossil fuels. While coal and oil are the most common fossil fuels, in some areas peat bogs provide a much less ancient fuel. Russia and Ireland produce and use great quantities of peat as fuel for generating electricity. These activities are certain to increase as world population grows.

A change in world climate is almost a sure thing if the atmosphere continues to accumulate carbon dioxide. It is called a greenhouse gas because it absorbs heat escaping from the earth and radiates it back like the glass panes in a greenhouse. Thus, the greater the concentration of carbon dioxide in the air, the warmer the world climate. The average global temperature today is 1°Fahrenheit (.55°C) warmer than it was a hundred years ago. The seven warmest years on record in more than a hundred years of record keeping have occurred since 1980.

Using the current rate of increase in atmospheric carbon dioxide, computer models predict the global temperature could increase by 4 to 9°Fahrenheit (2–5°C) by the year 2050. This would have disastrous effects on the world. Eventual melting of the polar ice caps would flood much of Florida, New York City, Los Angeles, and other coastal cities of the world. In the geologic past, rapid climatic changes have been associated with mass extinction of plants and animals.

The solution to the problem is easy to perceive but perhaps impossible to achieve. One obvious solution is to stop using so much fossil fuel. Trying to persuade the world to reduce the use of fossil fuels is such a complicated problem that there is currently no practical solution. Scientists estimate that to avoid climatic change, emissions of carbon dioxide must drop by at least 60 percent. This is not likely to happen. It may be that a warming world climate, with all it entails, is the price that must be paid for an exploding population and an energy-hungry civilization.

Water

Wetlands add water to the atmosphere in two ways, transpiration and surface evaporation. In those parts of the plant that are exposed to the air, wetland plants, like all terrestrial plants, have tiny openings in their leaves called stomata. Depending on the species, these range from 60,000 to 85,000 per square inch (12,000 to 17,000 per sq. cm) of leaf surface. Stomata evolved in response to an essential need of plants for the exchange of carbon dioxide and oxygen with the atmosphere. The inner spaces of leaves are always saturated with water vapor. When the stomata are open, water vapor diffuses from the leaf into the atmosphere. This is called transpiration, and there has been some debate among botanists as to whether it serves a useful function for the plant or is a necessary evil. The stomata must be open in order for plants to carry on photosynthesis. But the amount of water that plants lose into the atmosphere by this means is sometimes phenomenal.

The surface of wetlands varies from open water to saturated solid substrate. The maximum evaporation possible is constant and depends largely on climatic conditions such as temperature, relative humidity, and wind velocity. On a global scale, considering that only about 6 percent of the land surface is wetlands, they are responsible for a significant amount of gaseous water in the atmosphere. Although the extent to which wetlands influence regional weather has been incompletely investigated, there is little doubt that they do exert a considerable influence.

Plant Conservation

The Value of a Species

As members of the ecosystem, plants are producers of the food that is the base for all food chains. Without plants, all animal life would soon disappear. Each species of plant performs a unique function in its environment. If a species becomes extinct it leaves a hole in the ecosystem that can never be filled by another. The smallest flowering plant, duckweed, which looks like green pinheads floating on the water, is no less important than the tallest oak growing in the forest. They are each part of a web of food and energy that affects many other species. The survival of both is necessary to maintain an enduring natural system. It has been estimated that for each plant species that becomes extinct, an average of ten to thirty other species also disappear.

Habitat Destruction

Since colonial times, the non-native population of North America has grown from a few thousand to several hundred million. During much of this history, the prevailing attitude toward the environment has been governed by what ecologist Daniel Chiras has referred to as the frontier mentality. This attitude views natural resources as unlimited and existing solely for exploitation by humans. Humans are seen as masters of the natural world rather than as just another of its many species. Mastering the environment has become what amounts to an all-out assault with little concern for the consequences. In presettlement times, there was an estimated 215 million acres (87 million ha) of wetlands in the lower forty-eight states. About 56 percent of these have been lost since the late 1700s. Of the total number of threatened or endangered plant and animal species in the United States, more than half are dependent on wetlands for a portion of their life cycle. Until fairly recently, it seems as if man had declared war on wetlands, and even though that may have not been the intention, the effect is the same. In spite of a change in attitude—and in some states, laws protecting them—wetlands are still shrinking.

Habitat destruction in general is the main cause of extinction today. As a result of human activities, habitat destruction has been particularly intense in the past fifty years. Between 1850 and 1950, about eighty-six species of plants are known to have become extinct. Compare this with the period of 1973 to 1994, a spread of twenty-one years during which a hundred species may have been lost to extinction. During the same time, of the more than twenty thousand species of native vascular plants in the continental United States, about 750 had become endangered, which means they are in immediate danger of extinction. Another 1,200 are threatened and likely to become endangered in the near future. As many as one-third of these endangered and threatened plants are wetland species. This is not a problem restricted to the United States or North America. Of the more than 270,000 species of ferns, fern allies, and seed plants that are known in the world today, about 10 percent are estimated to be endangered, threatened, or rare.

Non-native Species

Habitat destruction is a broad problem with many facets. In addition to outright human dismantling, the introduction of plants from other lands, sometimes called alien or exotic species, can result in radical changes to an ecosystem. When plants from another ecosystem are introduced, they do not have natural enemies to limit their growth as do native plants. This often makes it possible for them to dominate their growing area to the exclusion of all other species. Native insects did not evolve with and are not adapted to use the introduced plants for food and shelter. As a consequence, alien species are frequently double threats to native ecosystems. They do not provide food and shelter for native species of animals, and they replace the plants that do provide these essentials.

Many foreign species arrived in North America by accident through international commerce and travel. Purple loosestrife (*Lythrum salicaria*, fig. 7.8) is a native of Europe that has invaded wetlands. Its beautiful pink to purple flowers in mid to late summer could be better appreciated were it not for the suspicion that the plant is displacing native species without performing their function in the chain of life. Water hyacinth (*Eichhornia crassipes*, fig. 7.2) in the southern states, a native of South America, and water chest-nut (*Trapa natans*, fig. 7.6) in the northern states, a native of Europe and Asia, are displacing native species of water plants and choking waterways. Regardless of the method by which alien plants arrive, many ecologists be-lieve that native ecosystems would be better off without them.

One further note on alien species is in order. While the above state-ments of traditional wisdom are very probably true, they may be oversim-plifications. The assumption is that if the ecological restraints of its home environment could be introduced, the alien nuisance species would no longer be a problem. This is no doubt true for many and perhaps for all ex-otics, but so little is known about the ecology of alien species that until a body of supporting data is available, it must remain an assumption. The statement that alien species do not provide food for native insects may also be questionable. For example, purple loosestrife, the very aggressive wet-land exotic, is insect pollinated. A brief observation will reveal that its flow-ers are visited for food by numerous insects, and it produces an abundance of seeds. It is not yet fully understood why some species grow so vigorously in environments away from their natural habitats. See chapter 7 for more information about alien species.

Wilderness Areas

There has been interest in maintaining wild areas in North America from as early as the mid 1800s. Naturalists such as John Muir, Aldo Leopold, and Robert Marshall have argued for the establishment of wilderness areas that would be protected from the influence of civilization. Such areas are more important today than ever because in competition with humans for space, wild plants and animals are losing on all fronts. Each species that is lost to extinction represents the loss of a potential life-saving drug, a source of food, or the loss of genetic material that could improve a crop plant. We still do not understand completely how an ecosystem works. It may be that wilderness areas will provide the natural laboratories that will yield infor-mation relative to human survival. But there is still another justification for setting aside protected areas. It is the recognition that plants and other ani-mals on the planet have as much right to be here as do humans.

A wilderness area must be large enough to include a functioning seg-ment of an ecosystem. For plants this means enough space to accommodate the life cycles of pollinators and for the operation of seed dispersal mechanisms. The Wilderness Act of 1964 set 5,000 acres as the minimum size and required that the area show no signs of human activity. Areas preserved under this act are mainly in the western United States and Alaska. The Eastern Wilderness Act of 1974 allowed smaller tracts to be set aside and permitted some signs of early human endeavors. About half of the land preserved under this act is in the Florida Everglades National Park and the Boundary Waters Canoe Area in Minnesota.

In protected wilderness areas, man is a visitor who can enjoy the beauty and the solitude but cannot stay. Theoretically his impact should be no more than any other large animal in the preserve. In some regions, so many people are finding the wilderness areas attractive that their wildness is being threatened. They are being threatened also by air and water pollution. However, their greatest long-range threat may be global warming brought on by the greenhouse effect.

Some Major Wetlands in North America

Atchafalaya Swamp

The Atchafalaya River (pronounced sha-fa-LI) is a distributary of the Mississippi River. This means that it is a branch that flows from the Mississippi rather than a tributary that feeds into the river. It flows for about 120 miles (190 km) through Louisiana to Atchafalaya Bay in the Gulf of Mexico. The lower two-thirds of the river meanders and branches through the third largest swamp in America. It has been estimated that at least 650 sq. mi. (1,700 sq. km) of swampland are maintained by the flow and flood of the river. The swamp is crossed by only one highway, Interstate 10, near its northern margin.

The Atchafalaya River is more than just a distributary of the Mississippi: it is the most direct route to the Gulf of Mexico. In the recent past, during a time of high water, it almost captured the main flow of the Mississippi. This would have been an economic disaster for Baton Rouge and New Orleans, since both of these cities depend on the Mississippi for ocean shipping. The Army Corps of Engineers built a lock and dam across the channel that connects the Mississippi to the Atchafalaya in an attempt to save the shipping route. The final word on these two rivers has yet to be written—the Atchafalaya may eventually have its way.

The vast acreage of this swamp is privately owned by lumber and oil companies. Most of the great stands of bald cypress and tupelo that once occupied large areas have been cut, but the lumber companies, waiting for another crop, have held on to the land. Only a few pieces of the swamp have been preserved for wildlife or for public visitation. They include a state wildlife management area near Grand Lake and Lake Pointe State Park and Longfellow-Evangeline State Commemorative Area, both near New Iberia, Louisiana.

Both man-made and natural influences have brought the future of the swamp into doubt. Sediments from normal flow, dredging, ditching, and flood waters are destroying as much as 25,000 acres (10,000 ha) of marsh and swampland each year. As water-carried silt is deposited during flooding, the bottom is built up and becomes dry land when normal flow returns. Some ecologists think the Atchafalaya Swamp in its present form is almost sure to be lost to future generations.

Cranberry Glades

Cranberry Glades is a small level valley of unique biological interest in the mountains of eastern central West Virginia. It consists of approximately 750 acres (300 ha) at an elevation of about 3,350 feet (1,022 m) surrounded by mountains that average more than 4,400 feet in height. It is an unusual wetland for several reasons. It is one of, if not the largest, peat bog area in the mountains of southern Appalachia. In eastern North America, it is rare to find peatlands as extensive as these at an elevation of over 3,000 feet. In addition, Cranberry Glades provides an environment where many northern species can survive far to the south of their normal range of distribution.

Unlike most peat bog localities of the northeast, the Cranberry Glades valley was not formed during the last glacial advance. Rather, the flattened area occupied by the glades is the result of lateral erosion of softer shales upstream from much harder rock formations. The main agents of erosion have been the Cranberry River, which bisects the area, and two tributaries, Charles Creek and Yew Creek. These two streams have divided the valley into four unequal sections with bog vegetation developing in each. An early study suggested the overflow during high water caused the development of levees along the banks of these streams. This created basins that subsequently became bogs (known in West Virginia as glades). Although the level valley may be a million or more years old, pollen studies from the deepest bog, called Big Glade, indicates that this bog began accumulating peat only about ten thousand years ago.

Perhaps the most interesting plant communities are found in the open glade areas or bog meadows. These provide habitats for both small cranberry (*Vaccinium oxycoccus*) and large cranberry (*V. macrocarpon*) for which the glades are named. In the past, cranberries have been harvested for sale and local use. If one is fortunate, on a trip to the bog in June or July, the flowers of native orchids rose pogonia (*Pogonia ophidglossoides*) and grass pink (*Calopogon tuberosus*) may be observed. A plant species with circumboreal distribution in more northerly climates, buckbean (*Menyanthes trifoliata*) blooms from April to July. Among the most fascinating bog species are the carnivorous plants. Round-leaved sundew (*Drosera rotundifolia*) and pitcher plant (*Sarracenia purpurea*) grow in the bog meadows. See chapter 5 for drawings and information about all the species named above.

Cranberry Glades serves as a refugium for many northern species. Dwarf cornel or bunchberry (*Cornus canadensis*), goldthread (*Coptis trifoliata*), and cranberries are species that normally grow much farther north. Buckbean and bog rosemary (*Andromeda glaucophylla*) occur in the glades in their southernmost-known colonies. Round-leaved sundew and the glades species of pitcher plant are northern species. Most of the above species are listed as rare plants in West Virginia. It is interesting to speculate on the methods by which these species arrived in this southern mountainous habitat. In all of them, the seeds are small with no visible adaptation for long-distance dispersal. It is known that the pitcher plant is not native to the glades but was transported to and planted in Big Glade in 1946. Since then it has spread and become thoroughly established.

In 1949 Cranberry Glades was designated a Natural Area of Monongahela National Forest. Today there is some concern for the survival of this unique landmark that the U.S. Department of Agriculture has referred to as a delicate botanical area. In times of heavy precipitation, this level valley acts as a catch basin for runoff from the surrounding mountains. In the past, construction projects and the accompanying forest denudation have resulted in soil erosion and increased sedimentation in the glades as streams overflow their banks. This may be the greatest threat to the acidic conditions necessary for maintaining a habitat suitable for rare bog plants. In 1972, tests of the substrate of the four bog communities show an organic matter content of only 10 to 23 percent. In most bog meadow communities the substrate is 80 to 100 percent organic matter. As with most bog communities, natural ecological succession will eventually convert the open glades to shrubs then to bog forest. The continued influx of mineral matter into the bog substrate will certainly hasten the transition to bog forest.

The Cranberry Glades Botanical Area is part of the 35,864 acre (14,346 ha) Cranberry Wilderness within the Monongahela National Forest. The Cranberry Mountain Visitors Center is located at the junction of Route 150 and Route 39/55. There is a half-mile-long boardwalk for visitor use that passes through two bog communities. Guided tours are conducted at 2:00 P.M. on Saturdays and Sundays throughout the summer months. Special tours can be arranged by contacting the visitors center. The bog community is a very fragile one. Visitors are urged to leave it undisturbed and view it only from the boardwalk.

The Everglades and Big Cypress Swamp

The Everglades was originally a single great, unobstructed river that flowed from Lake Okeechobee to Florida Bay, a distance of more than 100 miles (160 km). It is not a river in the sense that most people think because it averages only 6 inches (15 cm) in depth, is 50 miles (80 km) wide, and flows so slowly that the current is hardly detectable. During the dry season (winter), it may disappear entirely except for depressions and sloughs. Throughout much of its course, the riverbed is occupied by saw-grass (*Cladium jamaicense*, fig. 1.1), which is not really

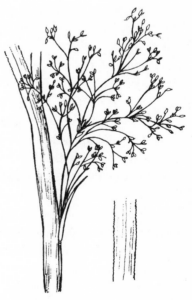

1.1. Saw-Grass (*Cladium jamaicense*)

a grass but a sedge. In reference to this growth, the Everglades has been called a "River of Grass."

The greatest problem facing the Everglades today is the lack of water. The drainage pattern from Lake Okeechobee has been so modified by human activities that Everglades plant and animal communities are in danger of extinction.

In 1947 an area of 1,301,349 acres (527,000 ha) was set aside as Everglades National Park. The park includes several habitats separated mainly by elevation. In the highest and driest parts are the pine lands, which are part of the southeastern evergreen forest. At the park entrance, the elevation is about 6 feet above sea level. As one proceeds farther into the park, the elevation drops. At about 4 feet above sea level is the river of grass called by the Seminole Indians pa-ha-okee, or "grassy waters." In this seemingly flat expanse are higher islands called hammocks that support tropical hardwoods and low spots that are occupied by bald cypress trees in clusters called cypress heads or cypress domes. At an elevation of 3 feet is the dwarf cypress forest. These are bald cypresses whose growth is stunted because of low levels of mineral nutrients in the substrate. At a still lower elevation, mangroves grow where freshwater mingles with the salty water of Florida Bay. The mangrove swamp forms a fringe around the southern and western tips of Florida.

The Big Cypress Swamp is adjacent to and west of the Everglades. Although bald cypresses are the dominant trees, the swamp is called Big Cypress because of it great size and not because of the size of the trees. This swamp, the Everglades, and the mangrove fringe add up to about 13,000 square miles (33,675 sq. km) of land. The physical conditions in Big Cypress are somewhat similar to those in the Everglades except the former does not depend upon the drainage from Lake Okeechobee. Instead, the Big Cypress Swamp is dependent on the approximately 50 inches (125 cm) of rainfall it receives during the rainy season (early June to late October). During the dry season, water stored in ponds and sloughs attracts and maintains animal populations until the return of the wet season. Corkscrew Sanctuary occupies 11,000 acres (4,452 ha) at the northern tip of Big Cypress Swamp. It contains the only stand of virgin bald cypress in North America.

The Great Dismal Swamp

The Great Dismal Swamp is located on the state line between southeastern Virginia and northeastern North Carolina. It has been under attack by man almost from the first settlement by white men. One of the early assaults on the swamp was by a company formed to drain 40,000 acres (16,196 ha) for growing cotton and rice. George Washington was a member of this company, and one of their canals still bears his name. The drainage efforts of this venture were successful, but the peat substrate contained so few mineral nutrients that farming was a failure. Following this attempt, Patrick Henry headed a group of business leaders who proposed a canal linking Norfolk Harbor in Virginia with Albemarle Sound in North Carolina. This project was completed in 1814 and exists today, paralleling U.S. Route 17, as the Dismal Swamp Canal.

The highest elevation in the swamp is along the western margin. It slopes gently eastward, and this is the general direction of water drainage. The source of water is a rock formation called the Norfolk aquifer along the western edge. Lake Drummond, just north of the center of the swamp, is a shallow lake with acidic, amber-colored water that seems to have originated about four thousand years ago. It has been suggested that the lake was formed either by a meteor collision or by a peat fire, but neither of these speculations have been confirmed by hard evidence. Peat underlies the entire swamp in varying thicknesses to almost 9 feet (3.3 m).

The original vegetation of the swamp consisted of stands of bald cypress, black gum, and Atlantic white cedar. It may have been the most northerly outpost of the great cypress-gum swamps. After logging and draining of the swamp, the type of forest that returned was very different. Red maple has become the most common tree and in areas of frequent fires, shrub growth has replaced the forest. In other sections, extensive tree farming has replaced natural vegetation. Many ecologists are of the opinion that the cypress-gum community of the swamp will never return.

The original size of Great Dismal Swamp was more than 770 square miles (1,994 sq. km). Human activities such as draining, ditching, logging, and fire have reduced the size of the swamp to only 330 square miles (854 sq. km). In 1973 it became the Dismal Swamp National Wildlife Refuge. Headquarters for the refuge are in Suffolk, Virginia. It offers 140 miles (225 km) of hiking and biking trails, a boardwalk, and boat access. Boat tours to

Lake Drummond are offered at the Dismal Swamp Canal Welcome Center on U.S. Route 17.

Okefenokee Swamp

The Okefenokee Swamp is located in southern Georgia on the Florida state line but most of the swamp is in Georgia. It is about 25 miles (40 km) wide, 40 miles (64 km) long, and covers an area of approximately 632 square miles (1,634 sq. km). Its origin goes back 250,000 years to a time when the shoreline of the Atlantic Ocean was 75 miles (120 km) farther inland and the region of the swamp was underwater. Wave action caused the formation of a sandbar east of today's Okefenokee. Then the land began to elevate and the shoreline receded, but the sandbar trapped water behind it forming a sandy-bottomed, shallow lake. Rainwater washed out the salt water and it became a freshwater lake. The sandbar is today known as Trail Ridge. This great reservoir of freshwater is the source of two rivers. The St. Marys flows to the southeast through Florida to the Atlantic. The Suwannee flows to the southwest through Florida to the Gulf of Mexico.

The Okefenokee supports a great variety of vegetation types, from bogs with native orchids and insectivorous plants, to marshes of submergent and emergent vegetation, to deep-water swamps of towering bald cypresses. These plant types grow on or around some seventy islands that cover 25,000 acres (10,123 ha). Some of these islands are floating vegetation mats similar to quaking bogs. This is the source of the name Okefenokee, which is a rough approximation of an Indian word that means "land of trembling earth."

After almost a century of exploitation by humans who cut its trees for lumber; shot, trapped, and poached its wildlife; and tried to drain its lifeblood, the Okefenokee National Wildlife Refuge was established in 1937. The protection of the swamp was further increased in 1974 when most of it became the Okefenokee Wilderness. Today there are only three entry points to the swamp: Okefenokee Swamp Park, a privately owned nonprofit organization near the town of Waycross at the northern end of the refuge; Stephen C. Foster State Park, near the center of the refuge but entered from the southwestern margin; and the National Wildlife Refuge Visitor Center at the Suwannee Canal on the southeastern margin. In the 1950s, the Okefenokee was brought to the attention of many people as the home of Walt Kelley's comic-strip character Pogo Possum. It was in these comics that Pogo uttered the ultimate environmental statement: "We have met the enemy and he is us."

The Pocosins

The Pocosins are shrubby, boggy, freshwater wetlands located on the Atlantic coastal plain from Virginia to northern Florida, and as far west as Alabama. More than 70 percent of these wetlands are in North Carolina. The Atlantic Ocean extended inland 100,000 years ago for a distance of over 100 miles (160 km) beyond the present shoreline, to a sand ridge

1.2. Fetterbush (*Lyonia lucida*)

called the Suffolk scarp. About eight thousand years ago, the level of the ocean dropped to approximately 400 feet below the present sea level. Since then the sea level has been rising. The land area between the older shoreline and the present sea level is relatively flat with many depressions. As the ocean level dropped, it was dissected into channels by ancient streams flowing to the sea. Peat began accumulating in these shallow depressions, then on the adjacent higher land eight to ten thousand years ago. Pocosin is an Algonquin Indian word that means "swamp on a hill."

Unique types of pocosin wetlands are the Carolina bays. There are five to six hundred of these scattered over the coastal plain. They are elliptical in shape, from .2 to .3 of a mile (.3–.5 km) in length, and have a southeast to northwest orientation. They have accumulated peat in thicknesses of up to 15 feet (4.6 m). The origins of the Carolina bays are unknown, but many geologists think they are the result of strong prevailing winds on early coastal water-filled depressions.

The pocosin peatlands are typically inhabited by evergreen shrubs with a scattering of trees that can survive in soil that is very low in mineral nutrients. The most abundant shrubs are leatherwood, fetterbush (*Lyonia lucida*, fig. 1.2), honeycup (*Zenobia pulverulenta*, fig. 1.3), and inkberry (*Ilex glabra*,

1.4. Inkberry (*Ilex glabra*)

1.3. Honeycup (*Zenobia pulverulenta*)

1.5. Loblolly Bay (*Gordonia lasianthus*)

fig. 1.4). The most common tree is pond pine (*Pinus serotina*). Other common trees, some of which may be stunted and appear to be shrubs, are red maple (*Acer rubrum*), sweet bay (*Magnolia virginiana*), loblolly bay (*Gordonia lasianthus* fig. 1.5), and Atlantic white cedar (*Chamaecyparis thyoides*). In areas of low shrubs and open spaces, insectivorous plants such as pitcher plants and sundews are frequently scattered.

Most pocosin plants have evolved features that promote survival following a fire. For example, the cones of pond pine remain closed for years, then open after exposure to the heat from a fire. This is a survival trait because the seeds germinate best on the mineral soil (ash) exposed after a fire. Although there is no evidence that pocosin shrub seeds require exposure to heat for germination, the establishment of seedlings rarely occurs in pocosins that have not been recently burned. Herbaceous plants like Vir-

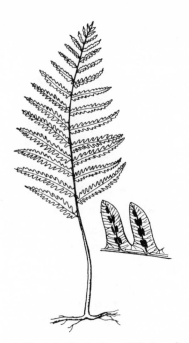

1.7. Giant Cane (*Arundinaria gigantea*)

1.6. Virginia Chain-Fern (*Woodwardia virginica*)

ginia chain-fern (*Woodwardia virginica*, fig. 1.6), giant cane (*Arundinaria gigantea*, fig. 1.7), and the sedge (*Carex striata*, fig. 1.8) are common in the pocosins. These plants usually do not produce seeds, but their most vigorous vegetative growth follows a fire.

Pocosin wetlands once covered about 247 million acres (100 million ha) in North Carolina. Drainage for agriculture, forestry, and peat mining had reduced this to about 694,000 acres (281,000 ha) by 1980. Today, undisturbed pocosin wetlands are continuing to shrink. There are sev-

1.8. Sedge (*Carex striata*)

eral places in North Carolina where they can be visited by the public. Nature trails, hiking trails, or guided tours are available at Carolina Beach State Park on U.S. Route 421 south of Wilmington; Croatan Forest south of New Bern; Jones Lake State Park in Bladen Lakes State Forest, southeast of Fayetteville; and Weymouth Woods Sandhills Nature Preserve on U.S. Route 1 near Southern Pines.

Prairie Pothole Marshes

The prairie pothole region is one of the largest wetlands in North America covering about 300,000 square miles (777,000 sq. km) in North Dakota, South Dakota, and Minnesota and the Canadian provinces of Manitoba, Saskatchewan, and Alberta. It consists of thousands of shallow ponds and lakes in depressions formed during the retreat of the last glacier. The rich soil and warm summers make this one of the most important wetlands in the world as a refuge for wildlife. At least 50 percent of all the waterfowl bred in North America in any year were hatched in this region. Unfortunately, since most of the region is privately owned farmland, drainage for agricultural use has reduced these wetlands to 40 or 50 percent of their original size. In response to alarm expressed by ecologists and sportsmen, state and federal governments and private conservation foundations such as the Nature Conservancy have initiated important protective measures. By the 1980s, the U.S. Fish and Wildlife Service had purchased or leased 1,500 square miles (3,885 sq. km) in North Dakota alone.

The vegetation of prairie pothole marshes goes through four recognizable stages over a five—to twenty-year period. The first is the marsh stage. During periodic droughts, the standing water dries up exposing the bottom mud. This allows emergent plants to germinate. Among these are cattails (*Typha spp.*), soft-stem bulrush (*Scirpus validus*), river bulrush (*Scirpus fluviatilis*, fig. 1.9), giant bur-reed (*Sparganium eurycarpum*, fig. 1.10), and common arrow-head (*Sagittaria latifolia*).

When rainfall and standing water return, the marsh exhibits a regeneration phase. For several years, the emergents increase in density and submergents sprout from seeds buried in the bottom mud. The submergents include pondweed (*Potamogeton spp.*), northern water nymph (*Najas flexilis*, fig. 1.11), hornwort (*Ceratophyllum spp.*), and the macroscopic green algae *Chara* and *Nitella* (fig. 1.12).

1.9. River Bulrush (*Scirpus fluviatilis*)

1.10. Giant Bur-Reed (*Sparganium eurycarpum*)

1.11. Northern Water Nymph (*Najas flexilis*)

1.12. *Chara sp.* (*left*); *Nitella sp.*

After a few years, for reasons not clearly understood, the emergents stop reproducing and a degeneration phase begins. Muskrat feeding and lodge building may contribute to this decline, especially with regard to cattails. Finally the emergents all but disappear, and the lake phase is initiated. The main vegetation during this phase consists of free floating duckweeds and submergents. The lake phase continues until another drought once again exposes the bottom mud.

Some potholes contain large boulders carried in by the glacier. An interesting explanation has been proposed as to why these are always in the deepest part of the pond. The boulders were used by buffalo to rub against in the shedding of winter fur. A smooth surface around the boulder at buffalo height supports this idea. As the animals rubbed against the boulder during droughts, their hooves trampled the soil into fine powder that was carried away by the wind. These deeper ponds are very important to the survival of some species during droughts today.

2

Types of Plants

For many people when they hear the word "plant," it brings to mind a tree, a house plant, a flower, or a weed. These are all plants but they are all of the same type. The seed-producing plants are the ones we most often think of because they are the largest and the most numerous plants on the earth. But there are other types of plants, and it would be difficult to explore any wetland without seeing some of them. This chapter examines the different forms of plant life that the naturalist is likely to observe in wetlands.

Algae

The many different kinds of algae are classified by color: blue-green, green, red, and brown. The color is caused by pigments that indicate how the sun's energy is used by that type of alga. They all have relatively simple structures and none of them have roots, stems, leaves, or flowers. The algae have been on earth for more than three billion years, and those ancient algae were the ancestors of all modern plants. The algae most likely to be seen in freshwater wetlands are the green and the blue-green. The blue-green algae are among the oldest photosynthetic organisms on earth and were probably the first to add oxygen to the atmosphere.

Even though they are less common than green algae, blue-green algae are widespread and commonly grow in streams, ponds, swamps, wet peat, and roadside ditches. They can carry on photosynthesis at a lower light intensity than green algae. This is a quality that home aquarium lovers may have noticed as the alga grows over the glass sides and sand of a neglected aquarium. Some species can thrive at high temperatures. They can be the most abundant organisms in hot springs that may reach temperatures of 163°F (73°C).

A feature of blue-green algae that is important in human cultures is that many species have the ability to convert atmospheric nitrogen into compounds called nitrates that can be used by all plants. Nitrogen is essential in plant metabolism for the manufacture of proteins. In India, the Philippines, and China, rice paddies have remained productive after centuries of rice cultivation by the inoculation of flooded rice fields with blue-green algae.

Some of the species of blue-green algae live in very close symbiotic associations with other species. One such relationship (discussed in a later section) is with fungi in lichens. Another is with the small aquatic mosquito fern (*Azolla spp.*, fig. 4.10). The alga grows and fixes nitrogen in cavities inside the fern leaf. Not only does *Azolla* provide large quantities of nitrogen to the soil, it also supplies compost or green manure that enriches the soil. The use of the water fern in rice fields may increase the rice crop yield by as much as 38 percent.

Blue-green algae are different from green algae in a number of ways. All algae have chlorophyll pigments that carry on photosynthesis, but the blue-green also contain a bluish-green pigment that gives them their characteristic color. They differ from green algae also in the arrangement of their genetic material. In green algae, as in most plants and animals, the genetic material (or DNA) in each cell is enclosed in a central structure called the nucleus. In the blue-green algae, there is no nucleus. Instead, the strands of DNA are dispersed throughout the cell. In this regard they have greater similarity to bacteria than to the green algae. Newer systems of classification reflect these similarities by classifying bacteria and blue green algae as cynobacteria and placing them in a separate group from other kinds of organisms called the Kingdom Monera.

The green algae are descendants of the blue-greens and are so-called because the main pigments that give them their color are green chlorophyll pigments. They grow in a wide range of habitats and are commonly seen as green slimy masses in roadside ditches, ponds, and streams. When examined under a microscope, these masses appear as delicate green strands, each consisting of many cells attached end to end. An alga often seen as a green scum on small ponds and roadside ditches is a species of the genus *Spirogyra* (fig. 2.1). In this genus, the structure that carries on photosynthe-

2.1. Spirogyra sp.

sis (the chloroplast) is in the shape of a spiral. The growth of either green or blue-green algae in water reservoirs can give a bad odor and taste to drinking water supplies.

Both blue-green and green algae reproduce asexually (without the union of male and female sex cells) by simple cell division and by fragmentation. In filamentous forms, each fragment develops by cell division into a separate strand when the filaments break. In the blue-greens, this is the only type of reproduction that has been observed.

Most of the green algae also can reproduce sexually. In the filamentous forms, the green masses that are visible to the naked eye are genetically haploid with only one set of chromosomes. Each species consists of two different strains that are often physically identical. One strain produces female gametes and the other male gametes. When two haploid gametes unite, a diploid cell called a zygote, with two sets of chromosomes, is produced. This is usually a thick-walled cell that may fall to the bottom of a pond or swamp in a resting phase that extends through the winter months. When the zygote germinates, it undergoes a special type of cell division called meiosis, or reduction division, in which the chromosome number is halved. The result is four cells that are haploid, each of which grows into a new filament.

Plant-like Organisms

Fungi

In older systems of classification, fungi were included in the plant kingdom. This may have been because they lack animal characteristics more than that they possess plant features. The fact is they have traits of both plants and animals. For example, they have rigid cell walls like plants, but like animals they do not possess chlorophyll or make their own food. However, fungi are a diverse group of organisms with characteristics that differ enough from both the plant and animal kingdoms to justify placing them in a separate category, the Kingdom Fungi. The ancestors of at least some of the fungi were probably green algae, and, like the algae, they have been on earth three billion years or more.

The growth form of fungi is basically filamentous consisting of long microscopic thread-like strands. A large number of these strands, called the mycelium, is usually dispersed in the soil or in the dead body of a plant or animal. Most fungi are saprobes, obtaining nourishment by secreting enzymes that digest organic material. At some point in the life cycle of many fungi, the strands of the mycelium grow together in a dense mass that appears aboveground as a macroscopic fruiting body. The function of the fruiting body is to form microscopic reproductive cells called spores. The spores are dispersed by wind, water, animals, and so on, and under adverse conditions they may remain viable for a long time. When they fall on a suitable medium, they germinate and grow into a new mycelium.

There are several types of fungi that commonly occur in wetlands. These are described below.

Sac fungi (Ascomycetes). These fungi are very important to humans in several ways. On the dark side, they are the causative agents in such diseases as athlete's foot, ringworm, and ergot poisoning, sometimes referred to as St. Anthony's fire. They are also responsible for many serious diseases of crop plants. On the bright side, the yeasts used in baking and in brewing alcoholic beverages are sac fungi. Others give the blue color to some cheeses and the distinctive flavor to Roquefort and Camembert, and they are the source of the antibiotic penicillin.

Some of these fungi have an aboveground fruiting body called an asco-

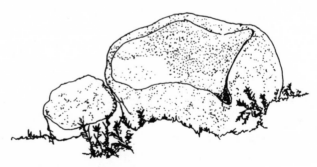

2.2. Blue-Staining Cup (*Caloscypha fulgens*)

carp. It is frequently shaped like a bowl or a concave disk. These range in size from less than $1/2$ inch to more than 5 inches in diameter. The inside lining of the bowl or disk may be a dull black or brown or it may be brightly colored yellow, orange, red, or purple. This col-ored layer contains asci or sacs in

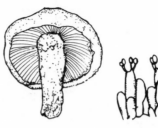

2.3. Gill Mushroom

which microscopic spores are produced that will grow into new underground mycelia. The sac fungi are sometimes mistakenly referred to as mushrooms, but real mushrooms are the fruiting bodies of another group, the club fungi.

Blue-staining cup (*Caloscypha fulgens*, fig. 2.2) is 1 to 2 inches (2–5 cm) in diameter with a cup of irregular shape. The outer surface is bluish-green to greenish with a bright orange-yellow interior. It can be observed in eastern North America from early to mid-summer. It grows in boggy areas but also in wet coniferous woods.

Club fungi (*Basidiomycetes*). These are probably more familiar to the general population than any of the other groups of fungi. They are called club fungi because the microscopic structures that produce spores, called basidia, are somewhat club-shaped. Two types of club fungi are commonly seen in wetlands, gill fungi and pore fungi. Gill fungi are common mushrooms that grow in a great range of sizes and colors. A gill mushroom consists of a stalk and an umbrella-like cap (fig. 2.3). On the underside of the cap, thin sheet-like gills radiate out from the central stalk. The spore-bearing basidia are located on each side of the gills.

The color of the spores is an important feature in the identification of mushrooms. Spore color can be determined by making a spore print. The stalk is cut off and the cap is placed gill side down on white paper and covered with a soup bowl or other convenient cover for a few hours. The spores will collect beneath the gills in enough quantity to form colored lines. To proceed further with identification, refer to the references at the end of this book.

Pore fungi do not have gills; the spore-bearing basidia are located in tiny tubes that open as pores on the underside of the fruiting body. These are best known as the bracket or shelf fungi that can be observed on dead trees, woody debris, or stumps in wetlands (fig. 2.4). The bracket fungi are important wood-rotting fungi that sometimes attack and kill living trees.

2.4. Bracket Fungi

Some of them cause dry rot and wood decay in homes and other wood structures. The fungal strands attack by digesting the cellulose in the wood. When the mycelium has become established in the wood, the shelf-like fruiting body develops and airborne spores are dispersed. Some of the bracket fungi are thick and woody with white undersides that are sometimes used as surfaces for artwork.

Water molds (Saprolegnia spp.). The water molds are aquatic fungi that grow mainly in freshwater swamps, ponds, lakes, streams, and in the surface layer of wet or damp soil. They may be saprophytic (live on dead organisms) or parasitic (live on living organisms) depending on environmental conditions and the susceptibility of the living hosts. They may attack fish, fish eggs, insects, and amphibians. These fungi are especially troublesome in crowded aquaria and in fish hatcheries. Water molds often start as wound parasites on adult fish, and infections have been observed in sport fish including Atlantic salmon, trout, and perch. Once an infection has begun on a fish, it attacks the scales, skin, and finally living tissue. The fungus grows rapidly and death may occur within twenty-four hours if it reaches the bloodstream.

Saprolegnia is not as commonly observed as sac and club fungi because it does not have a prominent macroscopic reproductive structure. When an organism has been infected, it is soon recognizable by the development of a white fuzz of mycelia around an entire small organism, such as an insect, or around the site of infection in larger organisms. *Saprolegnia* produces great numbers of motile reproductive spores in the water where it grows and in the effluent of sewage plants. See chapter 9 for an exercise on growing water mold.

Slime Molds (*Myxomycetes*)

The slime molds are so different that they may not even be related to the other fungi. They are sometimes classified as belonging to a separate kingdom, the Protoctista, which includes one-celled protozoans such as amoeba and paramecia. Instead of a filamentous structure, the slime mold is a macroscopic gelatinous mass of protoplasm called a plasmodium, with numerous embedded nuclei. Like a miniature version of a film monster, it flows very slowly along surfaces such as rotting logs on the edges of swamps or ponds, engulfing tiny bits of organic matter that it digests as food. In one phase of its life cycle, it forms stalked, often brightly colored sporangia in which hard-walled spores develop (fig. 2.5). These are dispersed by wind, and when one falls in a favorable environment it grows into a new plasmodium. Slime molds are most readily observed in moist to wet places where there is an abundance of decaying plant material.

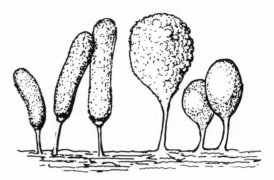

2.5. Sporangia of Slime Mold

Lichens

Lichens are everywhere—on rocks, on the bark of trees, on fence posts, on the stones or bricks of buildings, and sometimes even on sidewalks. They are everywhere, that is, except in areas with heavy air pollution. In the centers of heavy industrialization where sulfur dioxide pollution is greatest, there are practically no lichens. This dead zone extends outward for a considerable distance, especially in the direction of prevailing winds. Lichens, then, or their absence, are indicators of air quality.

A lichen is actually not one organism, but two. It consists of an alga and a fungus living in a very close association. The alga may be a green or a blue-green and the fungus is most often a sac fungus. The algal component manufactures the food while the fungus absorbs moisture and mineral nutrients. This type of association in which there are benefits for both organisms is called mutualism. Most of the body of the lichen is made up of compactly interwoven fungal strands. The alga forms a very thin layer just below the surface and constitutes about 5 percent of the dry weight. This relationship of alga to fungus is a very complex one that required many millions of years to evolve. Fossil evidence indicates the lichens first appeared on earth in the Mesozoic Era, which began about 225 million years ago.

Lichens are able to survive in very harsh environments. They grow on bare rocks where the temperature may reach 122°F (50°C) in summer and in the Antarctic where they survive and may even carry on photosynthesis at -65°F (-50°C). The surfaces on which they grow are usually nutrient poor, so their chief sources of mineral nutrients seem to be atmospheric dust and rainfall. Since nutrients are limited, the rates of growth are very slow, often less than one millimeter per year. Consequently, a large lichen is probably very old and some may be the oldest living things on earth. A few Arctic lichens have been determined to be 4,500 years old.

Three major growth forms of lichens are easily observed in the field: crustose, foliose, and fruticose. Crustose lichens are thin and very closely attached or embedded in the underlying surface (fig. 2.6). They can be seen on rocks but cannot be detected by touching. Some of the crustose forms are the most tolerant of air pollution. Because they may be present after other forms have disappeared from the area, they are an indicator species of polluted air.

Foliose or leaf-like lichens are thicker and usually have a central attachment to the substrate with unattached margins (fig. 2.6). During times of drought, the edges tend to roll up tightly and the lichen goes into a state of dormancy. With rainfall or an increase in humidity, it rehydrates and resumes photosynthesis. These and other forms of lichens are usually greenish-gray in color but in alpine conditions and in the Arctic they may be bright yellow, orange, or red. Most foliose and fruticose lichens produce special reproductive structures called soredia that are dispersed by wind.

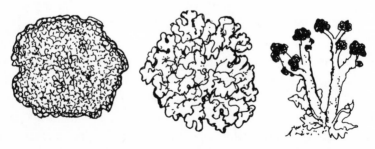

2.6. From left: Crustose Lichen; Foliose Lichen; British Soldiers (Fruticose Lichen)

Each soredium consists of a few fungal strands wrapped around one or more algal cells.

Fruticose lichens usually grow attached and perpendicular to the substrate, but some forms hang from tree limbs or other aerial perches. The fruticose forms are the most pollution sensitive of the lichens, and they are the first to disappear as air pollution increases. Familiar examples of these are reindeer lichen (*Cladonia rangiferina*), British soldiers (*C. cristatella*, fig. 2.6) and old man's beard (*Usnea spp.*). Reindeer lichens are common in the temperate zone in bogs, but they are also an important food source for Arctic animals such as caribou, musk ox and reindeer. Old man's beard is typically found in the southeastern United States in woody swamps and in areas where the average relative humidity does not usually drop below 47 percent. British soldiers are normally about 1 inch high with bright red tops.

Mosses and Liverworts

Mosses and liverworts are collectively known as the bryophytes. They are all small green plants whose ancestors were green algae. Mosses and liverworts are generally found in cool, moist habitats. They seem to be incompletely adapted to life on land because they lack many of the features that are necessary for successful survival out of water. For example, they do not have an efficient vascular system for absorbing and transporting water from the substrate to all parts of the plant. For this reason they are referred to as nonvascular plants. In complexity of the plant body and reproductive system, their evolutionary position is between algae and vascular plants such as ferns, gymnosperms, and flowering plants. However, bryophytes are not the ancestors of complex modern plants; they seem to be an independent line of evolution. The oldest known fossils of bryophytes are about 350 million years old, which is more recent than the oldest known fossils of vascular plants.

Bryophytes do not have true roots, stems, and leaves like vascular plants. Instead of roots, they have thread-like structures called rhizoids that anchor them to the soil and function somewhat as roots. Mosses and some of liverworts appear to have stems and leaves but without the conductive tissue of vascular plants. Some liverworts are not differentiated into leaf—and stem-like parts but consist of a flat ribbon-like structure called a thallus. It is usually fairly easy to distinguish between mosses and liverworts because mosses

2.7. Moss Gametophyte and Sporophyte

are erect and normally 1 to 3 inches high, while most liverworts grow flat against the ground. Individual bryophyte plants are small but typically grow massed in great clusters. Plants of this group have a worldwide distribution but are more abundant in temperate and arctic regions than in the tropics.

The life cycle of the bryophytes includes two genetically different phases. Organisms in the haploid phase have half the chromosome number, produce gametes, and are called the gametophyte generation. Organisms in the diploid phase have the full chromosome compliment, produce spores, and are called the sporophyte generation. Each of these phases is usually part of the plant body seen in the field.

The green leafy shoots of mosses are the gametophytes. These have sex organs at their tips that produce male and female gametes. When the sex organs mature, a motile sperm cell swims to and unites with an egg cell. The result of this union is a diploid cell, the zygote, which is the first cell of the sporophyte generation.

The diploid cell begins to divide immediately and forms a mass of cells embedded in the tissue of the gametophyte. These serve as the base for a long hairlike stalk that grows upward from the tip of the gametophyte. The stalk develops an enlarged capsule at its tip. The embedded tissue, the stalk, and the capsule make up the sporophyte.

In the field, a moss plant is usually observed as a green leafy basal part, the gametophyte, with the green or brown stalk of the sporophyte growing from its tip (fig. 2.7). The capsule is usually bent downward and often has a pointed cap that is the remnant of the female reproductive organ. It is called the calyptra, and it is helpful in the identification of some mosses.

Certain cells inside the capsule undergo a special kind of cell division called meiosis, or reduction division, in which the chromosome number is reduced by one-half. The haploid cells produced by this process are spores that are the first cells of the gametophyte generation. The spores are dispersed by wind, and when one falls on a moist substrate, it germinates and grows into a green algal-like filament. Buds develop along the filament and each bud grows into a green moss gametophyte. Individual filaments are very difficult to see, but they may occur in networks made up of many strands. A diligent observer may be able to see them as green thread-like lines near moss plants.

The mechanism for the dispersal of moss spores may be seen with a simple hand lens. An intact moss plant complete with gametophyte and sporophyte can be collected and taken indoors. The capsule of the sporophyte has a cap that covers a ring of enfolded flap-like teeth. If the moss plant is placed under an incandescent lamp, the capsule will dry, the cap will pop off, and the teeth will flip outward expelling the spores. When the capsule is remoistened, the teeth will return to their enfolded position.

The life cycles of liverworts are very similar to those of mosses except for the physical form of the sporophytes. In some of the thallose liverworts (those without leaf-like and stem-like structures) such as marchantia (*Marchantia spp.*), the sporophytes are much smaller and hang downward from special upright branches that look somewhat like miniature palm trees. In all instances the sporophytes of the bryophytes are more or less parasitic on the gametophyte generation. The bryophytes are similar to the vascular plants in exhibiting life cycles with alternating haploid and diploid generations. However, in all vascular plants the sporophyte is not parasitic on the gametophyte but is the most visible and dominant phase of the life cycle.

Economically the most important of all the bryophytes is sphagnum moss (*Sphagnum spp.*, fig. 5.1). The leaves of this moss contain alternating green living cells and dead hollow ones. These numerous hollow cells give the plant great capacity to absorb liquids. Absorbent cotton can absorb five or six times its weight of liquids, but sphagnum can absorb sixteen to eighteen times its weight. During World War I, it was used as a surgical dressing.

Sphagnum moss is a characteristic plant of peat bogs where it is harvested as the chief component of peat moss. Its great ability to hold water has contributed to the use of peat moss by gardeners as mulch to lighten the soil and keep it moist.

Ferns

The graceful beauty of ferns has always caught the fancy of humans and almost everyone can recognize a fern. This group of plants is diverse in size and growth form, and many do not fit the mold of what is thought of as a fern. For example, the tiny floating mosquito fern (*Azolla spp.*) may be less than 1/4 inch (6 mm) in diameter, and the cloverleaf fern (*Marsilea spp.*) looks like a floating four-leaf clover. At the other end of the size scale, some of the tree ferns in the tropics may be more than 50 feet (15 m) high. Although they have a worldwide distribution in a wide variety of habitats from the equator to the Arctic, 65 to 95 percent of all fern species grow only in the moist tropics.

According to the fossil record, ferns have been on the earth for about 370 million years. Their ancestors were land plants that evolved from green algae, although all the in-between stages are not fully understood. The ferns are a large successful group of plants, but they have never been a major component of world vegetation. They were abundant in the coal-age swamps and forests as they are today in the swamps and forests of eastern North America. They have never been the large conspicuous dominant plants, though, of any era.

Ferns and other plants with specialized tube-like cells for conducting water are called vascular plants. Most of these, including ferns, have well-developed roots, stems, and leaves. The ferns in the United States and Canada do not have erect aboveground stems but rather have underground ones called rhizomes.

Fern leaves are called fronds. They consist of a stalk or stipe that is attached to the underground stem, with an expanded portion called the blade. In a few ferns, the blade is undissected, forming a simple leaf, but in most ferns the blade is dissected one or more times forming the lacy leaf that is associated with ferns.

The fern leaf grows in a unique manner. Embryonic leaves that develop on the rhizome are very tightly coiled. As they mature, they unroll from the base in a growth pattern known as circinate vernation. Because the young uncurling leaves look like the neck and head of a violin, they are called fiddleheads. The unrolling of fern fiddleheads is always a welcome and attractive sign of spring. Fiddleheads of several species of ferns are collected in the spring and cooked as green vegetables. In some areas of Canada, fiddleheads of ostrich fern (*Matteuccia struthiopteris*) are harvested and frozen for commercial trade. These can be found in supermarkets in some American cities.

The life cycle of ferns includes a diploid sporophyte and a haploid gametophyte generation. Unlike mosses and liverworts, fern generations are completely independent of one another. The underground rhizome and the fronds make up the sporophyte generation. In some ferns such as cinnamon fern (*Osmunda cinnamomea*, fig. 2.8) and sensitive fern (*Onoclea sensibilis*, fig. 6.4), spores are produced on separate stalks, but in most species, the spores develop on the undersides of the fronds.

The underside of a frond may appear to be covered with brown dots. These are fruit dots, or sori, and each is a cluster of tiny, stalked, egg-shaped structures, the sporangia, that produce spores. Each sporangium contains special cells that undergo reduction division resulting in haploid spores with chromosome numbers half that of the sporophyte cells. Under dry conditions, the sporangium snaps open forcefully expelling the spores that are scattered by air currents.

2.8. Cinnamon Fern (*Osmunda cinnamomea*)

When a spore falls on a moist shady area, it germinates and grows into a very small, green, flat, heart-shaped structure known as the prothallus. On the underside of the prothallus are root-like outgrowths that attach it to the substrate and organs for the production of male and female sex cells. During a warm spring rain, or a warm night with heavy dew, a sperm cell swims to an egg cell fertilizing it and forming a diploid cell that becomes the first cell of the sporophyte phase. This cell develops into a leaf that grows upward and a root that grows downward, and the new sporophyte is established.

In earlier times, before their life cycles were understood, ferns were considered to be mysterious plants. Observers assumed they reproduced by seeds, but no seeds could be found. The brown dots on the undersides of the leaves were observed but no connection was made between these and fern reproduction, so the search continued for the illusive fern seeds. In those times the unknown was often associated with magic. This may have given rise to the legend that anyone in possession of a fern seed was invisible.

Fern Allies

Fern allies are the horsetails and club mosses. They are called fern allies because they have life cycles similar to ferns. They have actually been on the earth longer than ferns, and, like ferns, they are vascular plants whose ancestors evolved from green algae. Horsetails and club mosses have independent sporophyte and gametophyte generations. The visible green plants in these groups are the sporophytes; the gametophytes are small or underground and are very rarely seen. Today horsetails and club mosses are small plants 4 inches (10 cm) to 3 feet (90 cm) in height, but their ancestors were the most important trees of the coal-age swamp forests. They attained heights of up to 60 feet (18 m) or more and grew in vast swampy forests more than 300 million years ago.

Horsetails

The horsetails are all members of a single genus, *Equisetum*, that is frequently found in moist or wet open habitats. In field horsetail (*Equisetum arvense*, fig. 2.9), the sporophyte has two growth forms: a non-green spore-producing plant and a green photosynthetic one. The non-green cone-bearing branch arises from an underground rhizome in April. These have leafless jointed stems, are whitish-tan, and are 6 to 10 inches (15–25 cm) high. They shed their spores and wither in about two weeks. The green shoots appear a little later but do not reach their full development until late May or June.

The green phase of the field horsetail sporophyte consists of a central conspicuously jointed stem with branches in whorls. The jointed branches may have still smaller whorls of branchlets. The whole plant presents a bushy appearance that reminded someone in the past of a horse's tail. The leaves are very tiny and nonfunctional, but the green stem and branches

2.9. Field Horsetail (*Equisetum arvense*)

2.10. Scouring Rush
(*Equisetum hyemale*)

carry on photosynthesis. This plant lives for one season, then dies back to the underground rhizome in autumn. The rhizome is perennial and puts up new spore-bearing and green shoots each spring.

Scouring rush (*Equisetum hyemale*, fig. 2.10) is another common species of horsetails. It has an unbranched, jointed, green stem topped by an egg-shaped cone. The stem is evergreen, but new shoots appear each spring. The best development of cones is usually in June, but with some searching in a colony of scouring rushes, cones can be found even in winter. Scouring rushes usually grow in very damp or wet open spaces and are sometimes observed along railroad track clearings. The stems of both scouring rush and field horsetail have a high content of silica, the ingredient in sand. In former times, they were used for scrubbing pots and pans, hence the name.

Clubmosses

The clubmosses have been on the earth longer than either horsetails or ferns. The name clubmoss is misleading because they are not related to mosses and resemble them only faintly. There are several genera, but the one most likely to be observed in the field is *Lycopodium*. Most lycopods have horizontal stems that grow over the surface or underground with erect branches that seldom exceed 8 to 10 inches in height. These stems may branch repeatedly, with older portions dying, giving rise to several independent clumps of plants. In this way, large colonies will sometimes develop.

The lycopods are often found along the margins of swamps and bogs. Shining clubmoss (*Lycopodium lucidulum*, fig. 2.11) has upright branches 4 to 6 inches high with densely crowded leaves. Spores are commonly formed from July to September in sporangia located in the axils of leaves near the stem tips. During this same period small buds called gemmae develop at the stem tips. The buds may fall to the ground and grow into new plants.

In other clubmosses such as ground pine (*L. obscurum*, fig. 2.12) and ground cedar (*L. complanatum*, fig. 2.13), sporangia are clustered in cones at the tips of upright branches that usually shed spores in August and September. The spores are dispersed by air currents, and when they fall on suitable soil they have a tendency to sift downward to a depth of 1 to 4 inches. The spores germinate at this depth, and the gametophytes develop underground. The gametophyte is non-green and obtains food through a mutualistic or parasitic relationship with a soil fungus. It varies in size and shape and may be up to an inch in diameter or length. Sex organs develop on the surface of the gametophyte, and in a film

2.11. Shining Clubmoss (*Lycopodium lucidulum*)

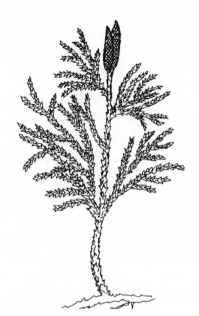

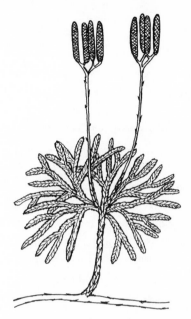

2.12. Ground Pine (*Lycopodium obscurum*)

2.13. Ground Cedar (*Lycopodium complanatum*)

of water the sperm swims to the egg to fertilize it, initiating a new sporophyte. The nature and location of sporangia, leaves, and horizontal stems are important features in the identification of the species of clubmosses.

Walking through a colony of clubmosses when the spores are mature will stir up clouds of yellow spores. They are produced in such great quantities that they are easy to collect. In colonial times, pioneer women used lycopodium spores as baby powder. Since they are water resistant, they have been used to coat pills to keep them from sticking together. Lycopodium spores are explosively flammable and have been used in fireworks and as flash powder for early photography. Clubmoss ancestors in the Coal Age also produced great quantities of spores. In fact, so many spores were released that in some regions, deposits of spores were converted to a special kind of coal called cannel coal. It is popular for fireplaces and campfires because it burns with a very bright flame and leaves a small amount of ashes.

The clubmosses are evergreen. This is unfortunate because it has resulted in their being collected to make Christmas wreaths. In colonial times this practice may have been acceptable because the human population was concentrated along the eastern seaboard and numbered in the thousands. Today it is unacceptable because with a human population numbering hundreds of millions it runs the risk of driving some species to extinction. The clubmosses are protected by law in several states and should be protected in all.

Gymnosperms (Conifers)

The gymnosperms are vascular plants that bear seeds in cones. The term *gymnosperm* means "naked seed" and refers to the fact that gymnosperm seeds are not enclosed in a fruit as are those of flowering plants. The gymnosperms have been on earth for about 325 million years. Their ancestors were vascular plants that evolved from green algae. They reached their peak of development during the time of the dinosaurs when they were the dominant vegetation on the earth. Since that time, they have declined as flowering plants have become the dominant vegetation.

Conifers

The most numerous and best known gymnosperms are the conifers. Trees of this type that are often found in wetlands include such familiar trees as pine (*Pinus spp.*), hemlock (*Tsuga spp.*),and spruce (*Picea spp.*). Although most conifers, including all of the above, are evergreen, two genera in North America, larch (*Larix laricina*) and bald cypress (*Taxodium distichum*), shed their leaves in autumn. A few conifers such as yew (*Taxus spp.*) are shrubby, but none are herbaceous.

The leaves of conifers are typically needle-shaped, as in pines, or they may appear to be overlapping scales as in Atlantic white cedar (*Chamaecyparis thyoides*). In pines and larch, most of the needles are attached in clusters on short spur branches. Pine needles are attached in clusters of two to five, while larch needles are usually shorter, softer, and attached in clusters of more than five. The needles of hemlock are flat while those of spruce have four edges, and they can be distinguished by rolling the needles between the thumb and forefinger. Although conifers produce a new crop of leaves each spring, all of last year's leaves are not shed. Leaves remain on the tree for one to five years and are shed gradually throughout the year. The shape, type of attachment, length, and texture of the needles are all important features in identifying species of conifers.

Like the ferns and nonvascular plants, the conifer life cycle alternates between a haploid gametophyte generation and a diploid sporophyte generation. Two types of cones are produced: male or pollen cones and female or seed cones. In the pollen cones, special cells undergo reduction division resulting in a great number of haploid spores that are shed as pollen grains. Each of these is a male gametophyte. All conifers are pollinated by wind, and in a coniferous forest, pollen grains are shed in such numbers that they can be seen as yellow clouds during peak periods of dispersal. A short time after pollen is dispersed, the male cones shrivel and fall from the tree.

The female cone is made up of flat scale-like divisions attached to a central column. On the upper surface of each scale, two mounds of tissue develop, and in each a single cell undergoes reduction division resulting in four haploid spores. Three of these disintegrate and the remaining cell grows into the female gametophyte bearing one or more egg cells. The pollen grain or male gametophyte, having been transported to the female cone by wind, develops a pollen tube that carries the sperm to the egg cell. The sperm and egg unite to form a zygote that grows into the young sporophyte. The embryonic sporophyte is surrounded by a layer of nutritive tissue and a hard outer covering. This is the seed that may become one of next year's crop of conifers. In most conifers, pollination usually occurs in May. In the temperate zone, the seeds of most species are shed in the autumn following pollination, but in pines they are shed in autumn the year after pollination. Each scale of the female cone bears two seeds with flat wing-like projections that aid in dispersal by wind.

Flowering Plants

The flowering plants are referred to as angiosperms, a term that means "seeds in a receptacle." This alludes to the fact that their seeds develop inside an ovary that matures to become the fruit. The flowering plants are the most recently evolved of all the plant groups. Their oldest fossils are about 130 million years old, and their ancestors were ancient gymnosperms that are extinct today. The time when dinosaurs roamed the earth was the age of gymnosperms, but today we are in the age of angiosperms. They dominate world vegetation with the greatest number of individuals and the greatest number of species. Unlike gymnosperms, they grow in a variety of forms including trees, shrubs, vines, herbs, and non-green parasites.

The flowering plants range in size from eucalyptus trees at over 300 feet (90 m) high and 60 feet (18 m) around at their base to some floating duckweeds that are 1/25 of an inch (1 mm) in diameter. They have covered the earth and have adapted to an amazing range of environments from the Arctic to the tropics. Some are water plants that grow completely submerged or in waterlogged soil. Others are drought-resistant plants that can survive the arid deserts of the world. Some species called air plants grow on the trunks of tropical trees with roots that hang freely and absorb water vapor from the air. Flowering plants dominate most of the terrestrial habitats on earth today.

Angiosperms are true land dwellers as are the mammals of the animal kingdom. In ferns, water is required for the sperm to swim to the egg in order to complete the life cycle. They resemble the amphibians of the animal world in this regard. In angiosperms, the link with aquatic ancestors has been severed. The male gametophytes or sperm cells are delivered to the egg cell by wind, insects, or some other animal pollinator; no water is necessary.

Mammals, birds, and insects have evolved in close association with flowering plants. The rise of herbivorous mammals was dependent on the development of grasses and other herbs. The birds have evolved with their main sources of food, the seeds and fruits of flowering plants or the insects that feed on leaves and fruits. The insects and the angiosperms have a remarkable history of codependence and coevolution. The domestication and subsequent cultivation of certain flowering plants was the initial step in the development of modern civilization.

The flower is the organ of reproduction for the angiosperms. A typical flower (fig. 2.14) consists of an outer ring of green leaf-like parts called the sepals. Collectively the sepals make up the calyx. Its function in the bud is the protection of the delicate inner parts. Inside the calyx is the corolla that is made up of individual parts called petals. The corolla in many flowers is brightly colored and associated with sweet-smelling nectar glands—features that attract insects and other pollinators. Inside the corolla is a ring of stamens, the male parts of the flower, each of which consists of a slender stalk, the filament, that supports the anther. In the center of the flower is the

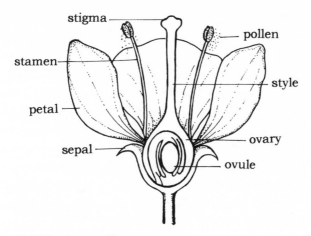

2.14. Angiosperm Flower

female reproductive structure, the pistil (there may be more than one), with an enlarged lower part, the ovary, an elongated neck, or style, topped by a sticky surface, the stigma, which receives the pollen grains. One or more ovules develop inside the ovary, and these will eventually become seeds.

As in other plant groups, angiosperms alternate generations as part of their life cycles. The anther contains four pollen sacs or sporangia that produce haploid male spores. These are shed as pollen grains that become the male gametophytes. Inside each ovule, there is a haploid female gametophyte consisting of only eight cells, one of which is the egg cell. During pollination, the pollen grain is transported to the stigma where it germinates. A pollen tube containing two sperm cells grows through the style to the egg cell. One of the sperm cells unites with the egg to form a diploid zygote that will become the next sporophyte generation. The other sperm cell unites with two other haploid cells and the resulting triploid cell grows into a food storage tissue known as the endosperm. At this point the ovule becomes a seed.

The stored food in the endosperm helps assure survival of the sporophyte seedling. Humans also have found this tissue to be an important source of food. It provides the nourishment of the cereal grains including wheat, corn, rice, oats, and rye. Coconut meat is endosperm tissue, and coconut milk is endosperm that did not form into cells. The food value of peas, beans, peanuts, and other legumes is endosperm that has been transformed and redeposited in two very thick seed leaves known as cotyledons.

Monocots and Dicots

There are two basic groups of flowering plants, the monocotyledons (monocots) and the dicotyledons (dicots), that are fairly easy to recognize in the field. Since there are twice as many dicots as monocots, the plants most often seen are dicots. Cotyledons are seed leaves, and as the names suggest, monocots have one and dicots have two. A seed usually consists of a seed coat, food storage tissue, and the embryonic sporophyte with its first leaves. Sometimes all the stored food is converted into the cotyledons as in beans and peas. The time to observe cotyledons is shortly after the seed germinates; the first structures to appear above the ground are the cotyledons, and whether there is one or two will be obvious.

There are features other than seed leaves that can be used in the field to distinguish between these two groups. The leaves of monocots have veins that are parallel from the base to the tip of the leaf; in dicot leaves, the veins are branched into a network. The most reliable and easiest way to distinguish between these two groups is by the number of flower parts. In monocots, the flower parts are in multiples of three. A typical species could have three sepals, three petals, and six stamens. Among wetland plants, common arrowhead is an example of this type. In some species, the sepals are modified so as to be indistinguishable from the petals; the iris flower appears to have no sepals but rather three outer arching and three inner erect petals. The monocotyledons include such familiar plants as grasses, irises, cattails, palms, and orchids. The orchids may be the second largest family of flowering plants and the most highly evolved of the monocots.

The flower parts of dicotyledons are in multiples of four or five. Almost all trees and shrubs and most herbs are dicots. A common flower type of these plants could have five sepals, five petals, and five to ten stamens. The aster family is the largest family of flowering plants and the most highly evolved of the dicotyledons. The flowers of this family are very small and so tightly clustered that each cluster gives the appearance of a single flower. For example, cutleaf-coneflower (*Rudbeckia laciniata*, fig. 7.11) looks like a flower with yellow petals and a dome-shaped, greenish center. Actually each of the "petals" is a flower, and the center is made up of many tiny flowers each consisting of a five-lobed corolla, five stamens, and a pistil with a curling two-part stigma.

3

Strategies of Survival

Genetic Variability

Asexual Reproduction

Most species of plants produce offspring by both asexual and sexual methods. In asexual reproduction, there is no union of male and female sex cells. Consequently the genetic makeup of the offspring is exactly the same as that of the parent. It can be as simple as a piece of the parent plant breaking off and growing into a new plant. This type of fragmentation is common in submerged water plants (hydrophytes) like algae and water-weed (*Elodea canadensis*, fig. 4.1). Another simple form of asexual reproduction is exhibited by common arrowhead (*Sagittaria latifolia*, fig. 6.14). It produces small tubers on its roots that will sprout into new plants in the next growing season.

In other species, asexual reproduction is a more structured part of the life cycle. For example, swamp loosestrife (*Decodon verticillatus*, fig. 3.1) grows along the margins of swamps, ponds, streams, and in shallow water. It has arching stems that take root and produce new plants where they touch the ground.

In the most complex form of asexual reproduction, seeds are produced without the union of male and female sex cells. Seeds are normally the result of sexual reproduction, but in sweet flag (*Acorus calamus*, fig. 6.9), an ordinary cell in the ovary begins to divide forming an embryo that will eventually become a seed. This type of nonsexual reproduction has the advantage of seed dispersal mechanisms for spreading into new areas. The seeds will germinate

3.1. Swamp Loosestrife
(*Decodon verticillatus*)

and, like other forms of vegetative reproduction, grow into a plant that is an exact genetic duplicate of the parent plant.

Maintaining an exact genetic composition has advantages in some situations. In harsh environments where there may be a close coordination between plant features and the demands of the habitat, a slight variation in plant form may result in extinction. Also, in rigorous arctic environments where there may be a scarcity of insect pollinators, asexual reproduction is sometimes the only road to survival. However, plant species that reproduce by asexual means only are evolutionary dead ends. In the absence of sexual reproduction, they cannot present a variety of genetic combinations for selection by a changing environment. Since they have only one combination, the only response they can make to climatic change is extinction.

Sexual Reproduction

In sexually reproducing species, there is a reshuffling of genes in each generation that provides a variety of genetic types. As a result, when there is a substantial change in the environment, although some members of the species will die, others may have genetic combinations that permit survival in the changed conditions. It is this quality that led to the origin of sexual reproduction and to its persistence in most living things today. All groups of plants including the algae, mosses and liverworts, ferns and fern allies, coniferous plants, and flowering plants have well-developed sexual systems. Since the flowering plants are the ones most commonly seen in the field, sexual reproduction in these will be discussed in more detail.

Methods of Cross-Pollination

Plants with the greatest genetic variability are best adapted for long-term survival. In flowering plants, the key to maintaining the greatest amount of variability is cross-pollination. Pollination is the transfer of pollen, the male sex cells, to the pistil, where the female sex cells are located. Most flowering plants have both male and female parts in the same flower. Cross-pollination occurs when the pollen from one plant is transferred to the pistil of another. When pollination is from the same plant, there is less variability in the offspring than when the pollen is from another plant. Numerous growth habits have evolved that promote cross-pollination.

Animal Pollinators

Insects

The most important agents of cross-pollination are insects. They have been associated with flowering plants for at least forty to sixty million years. During this time, many relationships have

3.2. Northern Blue Flag (*Iris versicolor*)

evolved in which plants and insects are mutually dependent on one another for survival. Most of the time, cross-pollination by insects is accidental. Visits to flowers are usually for nectar or pollen or both as sources of food. Insect-pollinated plants have sticky pollen that adheres to the body of the pollinator. When the insect visits another flower of the same species, it accidentally brushes against the stigma and the pollen is transferred.

Bee pollination. Bees are the most important of the insect pollinators. There are at least twenty thousand species, which all must visit flowers for food. Plant species pollinated by bees have evolved special types of flowers that are easy for bees to find and on which they can land. These insects cannot recognize the color red, but they can see ultraviolet light, which is invisible to the human eye. Flowers that have developed in response to bee pollinators are usually yellow or blue. Marsh marigold (*Caltha palustris*, fig. 6.17) has special markings visible only under ultraviolet light that highlight the location of nectaries. These are glands that produce a sweet substance called nectar that is used as food by many species of insects. Other bee flowers, like northern blue flag (*Iris versicolor*, fig. 3.2), have large lobes that serve

3.3. Tall White Orchid
(*Habenaria dilatata*)

as landing platforms and special markings called nectar guides that are like road signs directing bees to the nectar glands.

Some species of the orchid family have evolved highly specialized relationships with bees and bee relatives. In the bog orchid, grass pink (*Calopogon tuberosus*, fig. 5.15), the bees are duped by a cluster of yellow hairs on an almost vertical upper petal that are mistaken for yellow pollen-bearing anthers. After alighting on the petal, the bee discovers its mistake, but not before the petal collapses forward bringing the body of the bee in contact first with the stigma, then with the anthers of the flower. It then flies to a second flower and makes the same mistake, but this time it deposits pollen from the first flower.

Moth and butterfly pollination. Moths and butterflies are very important pollinators. Most moths are night flying creatures that have coevolved (developed in response to one another) with night-blooming plants. Since bright colors are not visible at night, most moth flowers are white or of a pale color that will stand out against a dark background. Moths have a well-developed sense of smell and flowers that attract them emit powerful fragrances after sunset. Both moths and butterflies have long sucking tongues that permit them to reach the nectar in narrow tubular flowers. Included among moth-pollinated flowers are the green woodland orchid (*Habenaria clavellata*) and the tall white orchid (*H. dilatata*, fig. 3.3), both of which grow in acid bogs.

Flowers pollinated by butterflies are usually showy, fragrant, and day blooming, as are the flowers pollinated by bees. Unlike bees, some butterflies can see the color red and visit red and orange flowers as well as blue and yellow ones. Some of the tropical and temperate zone milkweeds are pollinated by butterflies. The name of a native milkweed is swamp milkweed

3.5. Swamp Rose (*Rosa palustris*)

3.4. Swamp Milkweed (*Asclepias incarnata*)

(*Asclepias incarnata*, fig. 3.4). It has pink to rose-purple, crown-shaped flowers and it grows along the margins of open swamps. It is often visited by butterflies. One group of moths, the hawkmoths, are active during the day and may visit the same flowers as butterflies.

Fly and beetle pollination. The food of beetles and flies is frequently decaying fruit, dung, and dead animals. Plants that have been influenced by these insects in their evolution often have flowers with the unpleasant odors of rotting tissue. The sense of smell in beetles is more highly developed than the sense of sight and the flowers they visit are usually not brightly colored. Most beetles do not have mouth parts suited for obtaining nectar, especially from tubular flowers, so they feed on flower parts or pollen. There are at least 30,000 species of plants pollinated by beetles with more being discovered each year. Some plants visited by beetles are purple angelica (*Angelica atropurpurea*), fragrant water lily (*Nymphaea odorata*, fig. 4.8), and swamp rose (fig. 3.5).

A variety of flower types are pollinated by flies. Like beetles, they have

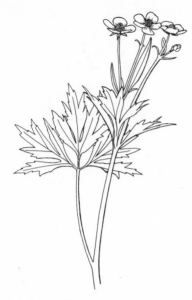

3.6. Swamp Buttercup (*Ranunculus hispidus*)

a highly developed sense of smell. Two plants that are commonly pollinated by flies are skunk cabbage (*Symplocarpus foetidus*, fig. 6.11) and swamp buttercup (*Ranunculus hispidus*, fig. 3.6). The flowering structure in skunk cabbage consists of many male and female flowers on a central column enclosed in a sheath. An odor similar to rotten meat attracts flies to this plant, and heat is generated by the central column that serves to intensify the odor. Other fly-pollinated species, like moccasin flower (*Cypripedium acaule*, fig. 5.17) and other lady's slipper orchids, are trap flowers which have odors that lure flies into a chamber of the flower. In order to enter, they must come into contact with the stigma, and they are dusted with pollen as they escape to visit the next flower.

Birds

In different parts of the world, many species of birds are specialized to feed on flower parts, flower-eating insects, or nectar. As with most insect pollinators, cross-pollination by birds is accidental. In North America, the most important bird pollinators are the hummingbirds, whose most important source of food is nectar. They have long slender beaks that can penetrate to the base of the longest tubular flowers. Hummingbirds have a well-developed ability to see color and can see the reds, but they have a very poor sense of smell. Consequently, plants pollinated by hummingbirds often have red flowers with little or no odor.

The red color of hummingbird flowers serves a dual purpose: it is a highly visible welcome mat for the birds and it discourages insects because they cannot see reds. Hummingbirds are heavier and require more energy to fly than insect visitors, hence the flowers they pollinate must produce large quantities of nectar. It is usually found in long tubular flowers or spurs

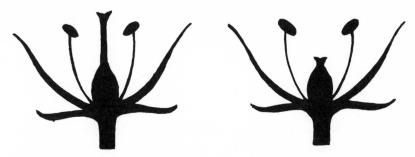

3.7. Type 1 flower (left); type 2 flower

that cannot be reached by insects. Even if insects could reach the nectar, they would be ineffective cross-pollinators of hummingbird flowers because the quantity of nectar would satisfy their need and they would not visit other flowers. Some typical common wetland hummingbird flowers are cardinal flower (*Lobelia cardinalis*, fig. 6.6), and touch-me-not (*Impatiens capensis*, fig. 6.18).

Genetic Safeguards

Even though cross-pollination by insects or birds is usually very dependable, there is still the possibility that pollen from the anther could reach the stigma in the same flower. This is called self-pollination, and many plant species have physiological or structural features to keep it from happening. One way has been through the development of different genetic strains within a species. The pollen of one strain is physiologically rejected by the stigma of any flower on the same plant. It must reach the stigma of another plant before it will grow a pollen tube that will result in the production of a viable seed. This is called self-incompatibility and is common among species of wetland plants.

A strategy to avoid self-pollination in some insect pollinated plants is the growth of pistils and stamens with different lengths. In some species, there are two types of flowers in approximately equal numbers: those with long-styled pistils and those with short-styled pistils (fig. 3.7). This greatly reduces the probability of pollen reaching the stigma in the same flower. However, as added insurance, this condition is usually accompanied by self-incompatibility. Fertile seeds can be produced only when pollen from a type 1 flower reaches the stigma of a type 2 flower. In purple loosestrife (*Lythrum salicaria*, fig. 7.8), there are two types of pollen, yellow and green, and three types of flowers with regard to length of pistils and stamens. Each type is self-incompatible.

Separation of the Sexes

As noted in the section on fly- and beetle-pollinated flowers, some plants have the stamens (male parts) and pistils (female parts) on separate flowers. These are called monoecious plants. Sometimes the male and female flowers are in different locations on the same plant as in giant bur-reed (*Sparganium eurycarpum*, fig. 1.10) and cattails (*Typha spp.*, fig. 6.15). In these species, the staminate (male) flowers are at the tip of the stalk, and the pistillate (female) flowers are beneath them. Other species that have separate male and female flowers on the same plant are the common arrowhead and European water-milfoil (*Myriophyllum spicatum*, fig. 7.4). Although self-pollination is less likely in these species, unless they are self-incompatible, it could occur.

In other species, the individual plants bear either male or female flowers but not both. These are called dioecious plants, and this arrangement is a guarantee that self-pollination can never occur. The disadvantage is that only half of the population can produce seeds since male and female plants are present in about equal numbers. Species with unisexual flowers include water-weed (*Elodea canadensis*, fig. 4.1), water smartweed (*Polygonum amphibium*, fig. 3.8), and false nettle (*Boehmeria cylindrica*, fig. 3.9). Plants with unisexual flowers, both monoecious and dioecious, are more common among wind-pollinated species.

Wind Pollination

Wind is an extremely inefficient agent of pollination. Whether or not an individual pollen grain reaches a stigma is purely a matter of chance. To ensure fertilization, wind-pollinated plants usually produce great quantities of pollen. So much is produced that even the most remote stigma is likely to be dusted. Only in this way can a seed crop big enough to sustain the species be assured.

Since their evolution was not influenced by animal pollinators, wind-pollinated flowers do not have colorful petals, do not produce nectar,

3.8. Water Smartweed (*Polygonum amphibium*)

3.9. False Nettle (*Boehmeria cylindrica*)

and do not have a fragrance. The stamens are usually long with anthers freely exposed to the air. The stigmas of wind-pollinated flowers are usually extensively branched to expose the maximum amount of surface. This increases their efficiency in trapping pollen grains. The ovary usually has only one ovule, so only one pollen grain is necessary for successful pollination.

Wetland species that are wind-pollinated include the cattails and giant bur-reed. These species grow on the edges of swamps and ponds or on damp to wet soil. Species that grow completely submerged must lift their flowers above the surface to achieve pollination. Among these are pondweeds (*Potomogeton spp.*) and the water-milfoils.

Water Pollination

Water as an agent of pollination is rare but can be observed in a few wetland species. In water-weed and water-celery (*Vallisneria americana*, fig. 4.3), staminate flowers develop underwater and then break off at maturity and rise to the surface. Pollen is released and floats to the stigmas of floating pistillate flowers. In water-celery, the pistillate flower is on a long stalk that re-tracts after pollination so the fruit develops underwater.

In hornwort (*Ceratophyllum demersum*, fig. 4.2) staminate flowers, there are twelve to sixteen anthers that break loose and rise to the surface at ma-turity. On the surface of the water, the anthers split, releasing pollen that sinks and comes into contact with the stigmas of the pistillate flowers.

Seed Dispersal

A discussion of seed dispersal cannot be complete without a clear under-standing of the relationship between fruits and seeds. Seeds develop within the ovaries of flowering plants, and the ripened ovary is a fruit. There are two types of ovaries, fleshy and dry. Fleshy fruits have thick walls that are sometimes colorful when the seeds are mature. Some of these are sweet and juicy and commercially identified as fruits when picked from trees or bought from the grocery store. Others, like tomatoes, cucumbers, and green peppers, are commonly called vegetables. Dry fruits are those in which the ovary wall is usually dry when the seeds are mature. In cultivated plants, beans and peas are of this type, as is swamp milkweed among the wild plants. In some dry fruits, the ovary contains only one seed, and at ma-turity the ovary wall becomes part of the seed coat. Members of the aster family have seeds that botanists classify as fruits.

Most seeds fail to complete their evolutionary mission: the growth of a new plant. Consider, for instance, bristly aster (*Aster puniceus*, fig. 7.9), a common plant in swamps, swales, and other moist places. Each plant has, conservatively, fifty to one hundred flower heads, and each flower head pro-duces from fifty to one hundred seeds. If every seed from every plant grew into a new plant, all wetlands would soon be overrun by bristly aster.

In spite of this high rate of failure, seeds continue to be the most im-portant means of reproduction and dispersal among flowering plants. They provide the plant with mobility, allowing the species to colonize new areas and increase its range of distribution. Still another advantage of seed dis-persal is that seedlings escape from competition with the parent plant. In response to these advantages, and perhaps others, a variety of seed dispersal mechanisms have arisen in plants. The two most common agents of disper-sal are wind and animals.

Dispersal by Animals

Animals are the most effective agents of seed dispersal. There are at least two reasons for this. First, migrating birds and mammals move at predictable seasonally regulated intervals. Over a long period of time, this could result in the development of plants with seeds that are mature during migration. Second, since animals usually move from one favorable environment to another, the seeds they are transporting are likely to be deposited in an area that is favorable for germination and growth. Dispersal is by three methods: (1) ingestion, (2) adherence to the outer surface of fur, feathers, or feet, and (3) transportation and storage as a food reserve.

Ingestion. Fleshy fruits have evolved chiefly as organs for seed dispersal. Animals are attracted to them as food sources, and seeds are transported in the intestines of the animals. Many seeds pass through animal digestive tracts unharmed. It is estimated that fruit eaters are responsible for seed dispersal in about 12.5 percent of the flowering plants in northeastern North America. Birds are the most important of these, but mammals and reptiles are also fruit eaters. In over 70 percent of the plants that have bird-disseminated seeds, fruit ripening coincides with the onset of fall bird migration.

Sometimes seeds pass through the digestive tracts of browsing or grazing animals. Such passage, whether in grazers or fruit eaters, often improves germination by softening hard seed coats.

Adhesion. Seeds dispersed in this manner are usually from herbaceous plants and have hooks, spines, or sticky surfaces. In bur-marigold (*Bidens cernua*, fig. 3.10), each flower head produces numerous seed-like fruits with four barbed awns that are very effective in dispersal by attaching readily to fur, feathers, and clothing.

3.10. Bur-Marigold (*Bidens cernua*)

Many aquatic plants produce small seeds with no apparent special structures for dispersal. The wide distribution of some of these species is the result of dispersal by water birds. Their adaptation for dispersal is small seed size: the seeds will stick to the mud on the feet of wading birds. The very rapid spread of the introduced plant pest purple loosestrife in eastern North America has probably been by this means. Dispersal by this method was recognized by Charles Darwin, who was probably the first to make a count of the seeds in the mud on the feet of wild ducks.

Dispersal by Wind

Wind is the second most important agent of seed dispersal. It is less efficient than animals for two reasons: (1) wind is highly variable and unpredictable, and it may not be present at the best time for dispersal, and (2) wind dispersal is random with many seeds falling in areas unsuitable for germination. Despite these shortcomings, species with seed modifications for wind dispersal are common. A few of these modifications are described in the following paragraphs.

Size. Very small seeds can be seen as an adaptation for dispersal by wind. The smaller the size, the greater the ease of dispersal. Members of the orchid family such as grass pink, moccasin flower, and rose pogonia (*Pogonia ophioglossoides*, fig. 5.16) have the smallest known seeds. One species has seeds that weigh 0.000002 grams each. It would take 500,000 of these to weigh one gram and 14,187,500 to weigh one ounce (28.4 g). These dust-like seeds can be widely dispersed in the slightest breeze.

Parachutes. A common adaptation for dispersal by wind is a tuft of hairs that function as a parachute. This type of dispersion is found in cattails, climbing hempweed (*Mikania scandens*, fig. 7.10),and bristly aster. Some species like swamp milkweed have seedpods that open to release seeds with parachutes of hair. About 16 percent of American plants are dispersed by seeds with parachutes.

Explosive Fruits

Probably the least effective method of seed dispersal, but a highly interesting one, is the explosive seedpod that flings seeds in all directions. This type of dispersal is exhibited by orange touch-me-not. It grows abundantly in eastern North America along the margins of swamps and in roadside ditches. At maturity, the seed capsule of this plant may be 1 1/4 inches (3 cm) long, and at the slightest touch, as it sways in the wind, it bursts open explosively. The capsule splits from the base into five segments that roll inward, forcefully hurling the seeds for a distance of 8 to 10 feet.

Adaptations for Water

The First Land Plants

The most radical change that ever happened to plant life on earth was the transition from water to a land environment. The first land plants evolved from green algae living on the ocean shore in the area between low and high tide, the intertidal zone, where they had daily periods of exposure to the air. This was a time when geological forces within the earth were causing the region to be pushed upward. As the intertidal zone was elevated, it first became an area of tidal pools. With further lifting, even these disappeared. During the millions of years required for this process, those species whose rate of evolution kept pace with the rate of shoreline elevation became land plants. Many species that were unable to make the transition to land became extinct.

The fossil record indicates that the first land plants may have resembled low-growing, branching, green sticks. In a later development, the ends of branches flattened and fused to form leaves. The plants that survived the rigors of natural selection had several structures that are necessary for survival on land. One was a root system that anchored the plant to the soil and absorbed mineral nutrients. Another was a waterproof, waxy covering, the cuticle, on aboveground parts that prevented extinction from excessive water loss. A third feature was a series of tiny closable openings in the cuticle, the stomata, that allowed an exchange of carbon dioxide and oxygen during photosynthesis. Finally, survival on land required a conduction system to transport water and mineral nutrients from the roots upward and food material from photosynthetic tissue downward to the roots.

With regard to their need for water, plants are classed as xerophytes (pronounced zerofites), mesophytes, or hydrophytes. On an environmental gradient of zero to ten, with zero representing the dry side, xerophytes are zero, mesophytes are five, and hydrophytes are ten. The first land plants to evolve were mesophytes, and these gave rise to the xerophytes and hydrophytes. The hydrophytes are the subject of this book.

Hydrophytes

Hydrophytes are plants that grow completely submerged, partially submerged, or with their roots in soil that is saturated with water for a portion of each year. Most mesophytes die of oxygen starvation if they are submerged or if their roots are underwater; water and water-saturated soil contain very low concentrations of oxygen. In the evolution of hydrophytes from mesophytes, several mechanisms developed in response to oxygen deprivation.

One of the characteristics shared by most hydrophytes is the presence of large cavities or air spaces in their tissues. These cavities serve two major functions: (1) they provide buoyancy that lifts the plant toward the surface and greater exposure to light, and (2) they act as gaseous reservoirs that accumulate carbon dioxide at night and oxygen during the day. Oxygen, the by-product of photosynthesis, collects during the day and is used at night when carbon dioxide is produced during plant respiration. This is particularly important to completely submerged species (submergents) that have no portion of the plant exposed to the air, like water-milfoil and pondweed.

Many species of hydrophytes have leaves that float on the surface of water or stems that extend above it. Their roots and the lower parts of their stems are underwater. The cavities in the tissues of these plants are interconnected, forming air passageways to all underwater parts. During photosynthesis, these passageways may become rich in oxygen, so the roots, even in the oxygen-free bottom mud, are not deprived. Species with floating leaves include fragrant water lily and water-shield (*Brasenia schreberi*). Among plants that extend above the surface of water (emergents) are the cattails and pickerelweed (*Pontederia cordata*).

In addition to having large air-filled spaces in their tissues, hydrophytes that are completely submerged have other features that are very different from mesophytes. They do not have waxy coatings or cuticles on plant surfaces, as do mesophytes, so the living cells are in direct contact with the water. In addition, leaves are usually very thin or finely dissected. In waterweed, the leaves are two cells thick. Water milfoil and hornwort have finely dissected leaves. As a result of these adaptations, submergents are capable of absorbing oxygen and mineral nutrients directly from the water rather than relying entirely on absorption through their roots. The main function of submergents' roots is anchorage.

Another feature of many submerged hydrophytes is nonfunctional stomata. Since these evolved in mesophytes as structures through which gases can be exchanged with the atmosphere, they obviously have no function in submergents. When they are observed in these plants, they are poorly developed and permanently open. Their presence is interpreted as supporting evidence that hydrophytes evolved from land-dwelling mesophytes. In emergent and floating hydrophytes, normally functioning stomata are present on the parts of the plants exposed to air.

3.11. Mermaid-Weed
(*Prosperpinaca palustris*)

In plants there is sometimes great variability or plasticity in the expression of genes. Although all the members of a species have very similar genes, two individuals grown in different environments may have quite different physical characteristics. This plasticity is dramatically demonstrated in some species of hydrophytes. White water-crowfoot (*Ranunculus trichophyllus*) and mermaid-weed (*Prosperpinaca palustris*, fig. 3.11) are sometimes emergents with leaves both above and beneath the surface of the water. The leaves below the surface are finely dissected like those of submergents while those above the water are not. Animals usually do not show this variability of genetic expression.

4

Marshes and Swamps

Wetlands are known by many names. One of the reasons is that wetlands are found in a great range of topographic conditions. Slight changes in topography can result in altered water levels and perhaps different plant species. One of the factors that defines a wetland, both ecologically and legally, is the type of plants it supports. Different topographic wetlands usually support different accumulations of plant species, and these often give them their names: a cattail marsh or a cypress swamp leaves little doubt about its identity. But wetlands are not always so clearly defined. Where their habitats grade into terrestrial ones, or into other wetland habitats, there may not be a clear demarcation of either a cattail marsh or a cypress swamp. For this discussion, marshes are considered to be wetlands that support only herbaceous plants and swamps as supporting both herbaceous and woody (trees and shrubs) vegetation.

Marshes are sometimes named for the dominant vegetation they support. Around the Great Lakes there are extensive (an estimated 1,400 marshes covering 465 square miles [1209 sq. km] in the United States and an even greater area in Canada) cattail (*Typha spp.*) marshes. Since cattails reproduce asexually by underground stems called rhizomes, these marshes are made up of large colonies of cattail clones. Another extensive freshwater marsh dominated by a single species is the great saw-grass marsh of the Florida Everglades (see chapter 1).

Other plants that commonly occur in freshwater marshes are grasses (*Poaceae*), sedges (*Cyperaceae*), and rushes (*Juncacae*). The plants in these families are very similar and all are often called grasses. For example, the "saw-grass" of the Florida Everglades marshes is actually a sedge. See chapter 9 for tips on distinguishing between the plants in these families.

Sometimes freshwater marshes cover large geographic areas. The prairie pothole region of the north-central United States and south-central Canada occupies 300,000 square miles (770,000 sq. km). This area was described in chapter 1. Wet meadows are types of marshes that occupy millions of acres in North America. These are areas that are seasonally flooded or in which the soil is saturated throughout the year.

The wetlands described in chapter 1 have been in existence for thousands of years. They appear to be in equilibrium with the environment, and some, like the prairie pothole marshes and the saw-grass marshes of the Everglades, show no signs of changing except where there has been human interference. They apparently will continue relatively unchanged as long as the current natural fluctuations in water supply are not diminished or exceeded. Over a long period of time, most wetlands change naturally through the process of ecological succession, or a natural replacement of one plant community by another culminating in one that reproduces itself and is not replaced. This self-sustaining stage is called the climax vegetation. In eastern North America, the final community is forest. It is not unlikely that marshes, in time, may become swamps, then swamp forests.

Swamps and marshes are commonly found along the margins of ponds and lakes where they form a transition zone between open water and dry land. If left to natural forces, ponds and lakes are geologically short-lived; they are destined to eventually be filled or drained. The filling process results from sediments that are washed in and from the accumulation of dead plants. The ways hydrophytic plants and ecological succession contribute to the filling of lakes, ponds, and wetlands are described below.

Submergents

The process of filling begins with microscopic one-celled plants, called phytoplankton, that die and settle to the bottom of ponds and lakes. This occurs very slowly, but gradually the water becomes shallow enough for light to penetrate to the bottom. Then submergent plants with roots attached to the bottom can become established from seeds carried in by streams or on the feet of waterfowl. Typical plants that may grow in deep water are water-weed (*Elodea canadensis*, fig. 4.1), water-milfoil (*Myrio-*

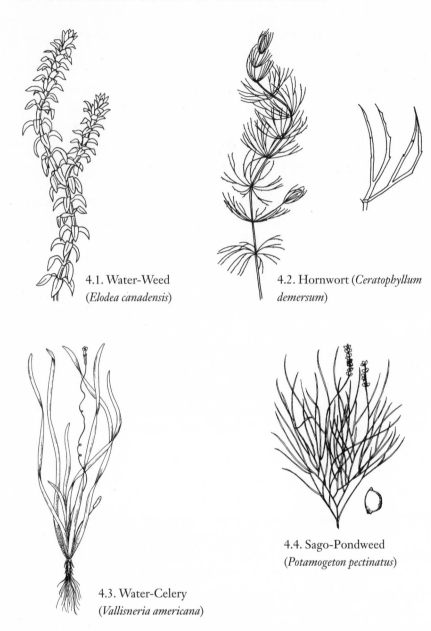

4.1. Water-Weed
(*Elodea canadensis*)

4.2. Hornwort (*Ceratophyllum demersum*)

4.3. Water-Celery
(*Vallisneria americana*)

4.4. Sago-Pondweed
(*Potamogeton pectinatus*)

phyllum spp., fig. 7.4), hornwort (*Ceratophyllum demersum*, fig. 4.2), water-celery (*Vallisneria americana*, fig. 4.3), and sago-pondweed (*Potamogeton pectinatus*, fig. 4.4). Each year some of these plants die and add to the bottom sediments.

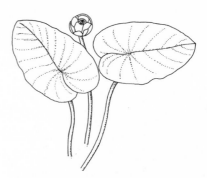

4.6. Yellow Water-Lily (*Nuphar variegata*)

4.5. American Lotus-Lily (*Nelumbo lutea*)

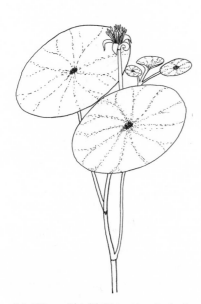

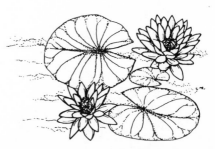

4.8. Fragrant Water-Lily (*Nymphaea odorata*)

Floating Plants

As the water depth decreases, some plant species are able to become anchored to the bottom by roots, with long stalks to leaves that float on the surface. Among these are American lotus-lily (*Nelumbo lutea*, fig. 4.5), yellow water-lily (*Nuphar variegata*, fig. 4.6), water-shield (*Brasenia schreberi*, fig. 4.7), and fragrant water-lily (*Nymphaea odorata*, fig. 4.8).

4.7. Water-Shield (*Brasenia schreberi*)

Other plants are free-floating without attachment to the bottom. These are mostly duckweeds, the world's smallest flowering plant. They

4.9. *From left:* Greater Duckweed (*Spirodela polyrhiza*); Lesser Duckweed (*Lemna minor*); Water-Meal (*Wolffia papulifera*)

4.10. Mosquito Fern (*Azolla caroliniana*)   4.11. Salvinia (*Salvinia rotundifolia*)

rarely flower, but even when they do the structures would not be recognized as flowers by most people. These plants reproduce asexually by producing buds on one end that detach and grow into new plants. Duckweeds are of three main types. Greater duckweed (*Spirodela polyrhiza*) is the largest and can be recognized by a cluster of two to sixteen roots that hang from the underside. Lesser duckweed (*Lemna minor*) is slightly smaller and has a single hanging root. Water-meal (*Wolffia spp.*) resembles a green dot on the water, is about $1/_{25}$ of an inch (1 mm) in diameter, and has no roots (fig. 4.9).

Some species of aquatic ferns are free floating. One of the most common is mosquito fern (*Azolla caroliniana*). It is somewhat larger than greater duckweed at $1/_5$ to $2/_5$ of an inch (5–10 mm) in width with scale-like overlapping segments (fig. 4.10). Another free-floating fern more common in southern waters is salvinia (*Salvinia rotundifolia*, fig. 4.11). It has oblong oval-shaped leaves that are $^1/_2$ to $^3/_4$ inches (12–19 mm) long.

Other floating plants that are sometimes a nuisance because they clog waterways with excessive growth are water hyacinth (*Eichhornia crassipes*, fig. 7.2) and water chestnut (*Trapa natans*, fig. 7.6). Water hyacinth is a native of tropical America that was accidentally introduced into Florida waters. It is now widespread in the southeast as far north as Virginia and Missouri. Water chestnut is more common in northern waters. It is a native of subtropical Africa and Eurasia. Early in its development it is attached to the bottom by a long stalk, but this frequently breaks and the plant becomes free floating.

Emergents

As floating plants die and add to the bottom sediments, the water depth continues to decrease. Eventually plants can become established that are rooted on the bottom with stems and leaves above the surface. Some of the most common emergents are pickerel-weed (*Pontederia cordata*, fig. 4.12), common cattail (*Typha latifolia*) and narrow-leaved cattail (*T. angustifolia*, fig. 6.15), arrow-arum or tuckahoe (*Peltandra virginica*, fig. 4.13), common

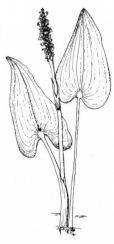

4.12. Pickerel-Weed (*Pontederia cordata*)

4.13. Arrow-Arum (*Peltandra virginica*)

4.15. Sandbar Willow (*Salix exigua*)

arrowhead (*Sagittaria latifolia*, fig. 4.14), softstem-bulrush (*Scirpus validus*, fig. 4.14), and olney-threesquare (*S. americanus*).

Shrubs

With time, the emergents add to the bottom accumulation, making it shallow enough for woody plants to gain a foothold. Water-loving shrubs or trees may begin to grow along the margins and extend inward on the shallowest areas. This stage of swamp development may persist for many years if the water level does not change. Some examples of common shrub and swamp forest plants are given below.

Sandbar willow (*Salix exigua*, fig. 4.15) is a shrub with multiple stems that may grow to 9 feet (3 m) in height. It is dioecious with male and female flowers in catkins. With smooth reddish-brown branchlets, it has leaves that may be up to 5¹/₂ inches (14 cm) long. In all wil-

4.14. Softstem-Bulrush
(*Scirpus validus*)

4.16. Speckled Alder (*Alnus incana*)

4.18. Buttonbush (*Cephalanthus occidentalis*)

4.17. Red Osier-Dogwood (*Cornus sericea*)

lows, the buds are covered by a single hollow conical scale instead of overlapping scales as in most other trees and shrubs.

Speckled alder (*Alnus incana*, fig. 4.16) is a shrub or small tree that may be up to 20 feet (6 m) in height. It is dioecious with male and female flowers in catkins. The common name refers to spots on the bark that are modified stomata called lenticels. It is circumboreal in distribution and extends as far south as West Virginia. Speckled alder is wind pollinated and produces numerous tiny, slightly winged seeds that are dispersed by wind and water.

Red osier-dogwood (*Cornus sericea*, fig. 4.17) may grow to 10 feet (3 m) in height sometimes forming dense thickets. The twigs are bright red with large white centers. It has small, white, insect-pollinated flowers in flat or convex terminal clusters that become clusters of white berry-like fruits. Red osier-dogwood is cultivated extensively as an ornamental and is easily propagated by cuttings.

Buttonbush (*Cephalanthus occidentalis*, fig. 4.18) is a shrub or occasionally a small tree that may grow to 10 feet (3 m) in height. Leaves are opposite or sometimes in whorls of three or four, oval shaped with smooth margins. It has small, white, insect-pollinated flowers, with styles extending far beyond the petals, in dense spherical clusters at the tips of stems and branches. Buttonbush has a very wide geographic distribution ranging from

4.19. Sweet Bay (*Magnolia virginiana*)

4.20. Red Maple (*Acer rubrum*)

Nova Scotia and Quebec to Minnesota and south to Mexico and the West Indies.

Sweet bay (*Magnolia virginiana*, fig. 4.19) is a tall shrub or slender tree that grows to 65 feet (20 m) in height. The leaves are shiny green, oblong, aromatic, and 7 inches (18 cm) long. It has very fragrant white flowers about $2^1/_2$ inches (6 cm) wide. Sweet bay is more common in southern states where the leaves are evergreen, but it extends along the coastal plain to Long Island, New York.

Red maple (*Acer rubrum*, fig. 4.20) is a monoecious deciduous tree that may grow to 117 feet (35 m) in height. It has reddish buds, red flowers, red leaf stalks, red autumn foliage, and reddish fruits. Pollination is by wind, and the winged fruits are dispersed by air currents. The leaves are oppositely attached with three shallow lobes and usually two smaller lobes near the base. Red maple is a common tree from southeast and south-central Canada to Florida and Texas. Other common names that have been applied are swamp maple, white maple, soft maple, and shoe-peg maple.

Green ash (*Fraxinus pennsylvanica*, fig. 4.21) is a dioecious or monoecious broad-leaved tree, with opposite leaves, that may reach 60 feet (18 m) in height. The leaves are pinnately compound, usually with seven leaflets. The female flowers are wind pollinated and the fruits are winged for dispersal by air currents. It is probably the most widespread of the ashes ranging from Nova Scotia to Alberta, south to Florida and Texas. A northeastern variety with hairs on the under surfaces of leaves is called red ash. Other

4.21. Green Ash (*Fraxinus pennsylvanica*)

4.22. American Elm (*Ulmus americana*)

4.23. Sweet Gum (*Liquidambar styraciflua*)

names that have been used for green ash are water ash, river ash, and swamp ash.

American elm (*Ulmus americana*, fig. 4.22) is a large tree that may reach 130 feet (39 m) in height with a trunk diameter up to 5 feet (1.5 m). Alternately attached leaves have a slightly roughened upper surface and a double toothed margin. The flowers appear before the leaves and have both stamens and pistils. They are wind pollinated and produce fruits about $^1/_2$ inch (1.3 cm) in diameter with a winged margin for dispersal by wind. This beautiful shade ornamental of the past is rarely seen today as a large tree because of its destruction by Dutch elm disease. Smaller trees to 25 feet (7.5 m) high are still present in swamp forests. American elm is also known as white elm.

Sweet gum (*Liquidambar styraciflua*, fig. 4.23) is a tall monoecious tree that may grow to 140 feet (42 m) in height. The leaves are alternately attached with long stalks, and five to seven pointed lobes give them star shapes. The female flowers develop in spherical burr-like structures about $1^1/_2$ inches (3.7 cm) in diameter, on long stalks that remain on the tree

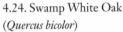

4.24. Swamp White Oak
(*Quercus bicolor*)

4.25. Black Gum (*Nyssa sylvatica*)

throughout the winter. Beginning in autumn, numerous winged seeds are shed and dispersed by air currents. It is sometimes the source of a wax called storax or styrax that resembles beeswax and is used in pharmaceutical preparations, incense, perfuming powders and soaps, and flavoring tobacco. During autumn, the leaves turn a brilliant red. Sweet gum ranges from southern Connecticut to southern Illinois south to Florida and Guatemala. It is also known as bilsted and red gum.

Swamp white oak (*Quercus bicolor*, fig. 4.24) is a large tree, with sometimes drooping branches, that may reach 70 feet (21 m) in height. The leaves are alternately attached, widest above the middle with blunt tips and shallow rounded lobes. They are green and somewhat shiny on the upper surface with hairs on the lower surface giving them a whitish cast; thus the species name *bicolor*. This is a monoecious tree with male flowers in catkins and female flowers usually producing pairs of acorns on stalks longer than the leaf stalks. It has a higher tolerance for ozone than most other swamp forest trees. Swamp white oak ranges from Maine and Quebec to Minnesota, and south to North Carolina and Arkansas.

Black gum (*Nyssa sylvatica*, fig. 4.25) may grow to a height of 100 feet (30 m) with a trunk diameter of up to 3 feet (1 m). It has leaves that are alternately attached, oval shaped with pointed tips, shiny green, leathery, crowded toward the ends of twigs, and turning bright red in autumn. The tree is bee pollinated with male and female flowers usually on separate trees. Female flower clusters usually produce one or two single-seeded, blue-black, berry-like fruits. Black gum is one of the few trees that can survive long periods of flooding. Other names that have been used for it are sour gum, water tupelo, swamp tupelo, and black tupelo.

The roots of most of these trees do not penetrate deeply into the substrate since water-logged soil is very low in oxygen. Instead, they develop shallow root systems and may be uprooted in high winds. A period of several years when there is high water during the growing season may kill many of the trees and return the swamp to shrub or emergent vegetation. The swamp forest stage may persist for a very long period of time if regional climatic conditions do not change.

Sometimes along the margin of a pond or lake the wetland successional stages can be observed as distinct zones of vegetation. There is evidence that this sequence of stages takes place in many but not all wetlands. The order of occurrence of the stages may not be as described above. For example, floating plants, emergent plants, and shrubs may appear in a single zone. It is also known that some wetlands are thousands of years old but show no signs of being replaced by woodlands.

All swamp forests do not originate as progressively filled marshes or swamps. In some lowlands and river floodplains, periodic flooding is a normal environmental fluctuation.

The exact species composition of swamp forests in Mississippi will not be the same as those in New York State or Ontario, Canada, but some of the species described above will be found in all three. In drier areas of swamp forests, they usually grade into upland mesic deciduous forests. At the other end of the water gradient are deep-water swamps. These are discussed in the following section.

Southern Deep-Water Swamps

Southern deep-water forested swamps have been described as the most enchanting wetlands in North America. They have become familiar to many Americans as settings for movies and dramatic television productions. They are indeed fascinating, if sometimes forbidding places, with tall trees arising from dark waters of unknown depths. Part of the feeling of mystery and enchantment, or dark and threatening gloom, is provided by gray-green Spanish moss (*Tillandsia usneoides*, fig. 4.26), which dangles from branches, swaying with the softest breeze. Spanish moss is not a moss at all but an air

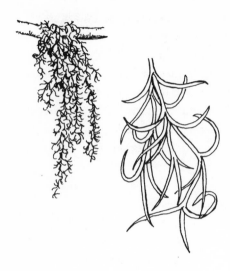

4.26. Spanish Moss (*Tillandsia usneoides*)

4.27. Bald Cypress (*Taxodium distichum*)

4.28. Water Tupelo (*Nyssa aquatica*)

plant. It is a flowering plant of the pineapple family that is able to get mineral nutrients from atmospheric dust and water by absorbing gaseous water from the air.

Forested swamps occur along the Atlantic coastal plain from Delaware to Florida, along the Gulf Coast to eastern Texas, and on the flood plain of the Mississippi River as far north as southern Illinois. These swamps have been reduced to 40 percent or less of the area they occupied in the early 1800s. They have been heavily cut over for valuable lumber and drained to create croplands. Fortunately a few of them are difficult to reach for lumbering, with the result that some of the most prime wetland forests in the world have been preserved.

The dominant trees in many of the swamps that have standing water for all or most of the year are bald cypress (*Taxodium distichum*, fig. 4.27) and water tupelo (*Nyssa aquatica*, fig. 4.28). Bald cypress is a gymnosperm, but it is not evergreen. It sheds its leaves every year as do deciduous trees. It is a very large tree that may grow to 130 feet (40 m) in height. A smaller variety called pondcypress grows mainly in Georgia and Florida, but its range extends northward along the coastal plain to North Carolina. The needles of bald cypress stand out in a flat plane at right angles to the branch. In pond-cypress, the needles are sometimes wrapped around the twig in an overlapping pattern. Bald cypress sometimes grows in pure stands as in Big Cypress Swamp of southwestern Florida. An excellent untouched cypress swamp is Florida's Corkscrew Swamp. It can be explored by a boardwalk that winds through its watery interior, a particularly interesting trip during the summer and fall wet season.

Water tupelo or cotton gum is a large deciduous tree that may reach a height of almost 100 feet (30 m). It sometimes grows in pure stands but more often in a mixture with bald cypress. These mixed stands may form a canopy so dense that transmitted light is not enough to support the growth of understory plants. If the canopy is less dense, the submerged, floating, and emergent vegetation described earlier may thrive in the water of the swamp.

Other trees that sometimes grow with cypress and water tupelo are water elm (*Planera aquatica*, fig. 4.29), green ash, and black gum. In winter when the large trees lose their leaves, a touch of color is retained in the swamp by evergreen shrubs and small trees such as sweet bay, leatherwood (*Cyrilla racemiflora*, fig. 4.30), and wax-myrtle (*Myrica cerifera*, fig. 4.31).

Two features characteristic of trees that grow in southern deep-water swamps are knees and enlarged bases (fig. 4.32). Knees are pointed projections of the roots that normally rise above the water level. In cypress swamps they are usually no more than 3 to 4 feet (1–1.3 m) in height, but in

4.29. Water Elm (*Planera aquatica*)

4.30. Leatherwood (*Cyrilla racemiflora*)

4.31. Wax-Myrtle (*Myrica cerifera*)

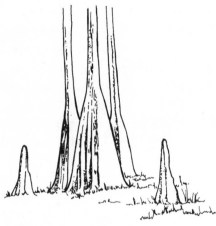

4.32. Buttresses and knees

some instances they may be 9 to 12 feet (3–4 m) high. Their function is not clearly understood. They may be a means of supplementing the oxygen supply to the roots, but when they are removed it does not seem to harm the tree. Since the knees have a root system of their own it has been suggested that they may help anchor the tree. Although knees are most abundant on cypress trees, they are also produced by water tupelo and black gum when these trees grow in standing water.

Enlarged bases or buttresses develop on cypress, water tupelo and black gum. The enlargements may extend upward from the base of the trunk from three to several feet (dm). The height of the buttress depends on the height to which the tree is constantly or regularly underwater. Since the enlargement usually extends well above the water line, it has been suggested that the buttresses may assist in providing oxygen to the roots. The enlargements may also contribute to the tree's stability in the water-soaked substrate. Aside from these speculations, the functions of the buttresses are unknown.

The seeds of both water tupelo and bald cypress require oxygen and moist soil for germination. Thus, if the deep water swamp does not occasionally go dry it will become open water as the trees die. However, these trees live for hundreds of years, so in order to maintain the swamp forest, drying would need to occur only once every two or three hundred years.

Bogs and Peatlands

Uses of Peatlands

Definitions

There are two types of peatlands that are differentiated by their water supply. Those receiving water that has percolated through soil and is rich in mineral nutrients are called fens. Those in which the main source of water and mineral nutrients is rainwater are referred to as bogs. In northeastern North America, the latter are usually the result of glacial damming or the burial of large chunks of ice. When the ice melted, the water bodies eventually became bogs. They are sometimes called raised bogs.

The distinction between bogs and fens is not always clear in the field because the same species of plants may occur in both. To simplify terminology, bogs are peatlands that develop in discrete basins and support typical bog vegetation. Organic substrates that cover large areas not confined in specific basins will be described by the more general term peatlands. Peat is incompletely decomposed plant matter formed in deposits that are saturated with water. All undisturbed peatlands are characterized by an organic substrate usually covered by a growth of sphagnum moss (*Sphagnum spp.*, fig. 5.1).

Sources of Food

Peatlands have been used by humans for thousands of years and continue to have great value in our culture. One of the earliest uses, and one that con-

5.1. Sphagnum Moss
(*Sphagnum* sp.)

tinues today in some regions, is as a source of food. Highbush blueberry (*Vaccinium corymbosum*, fig. 5.2) grows in bogs from Nova Scotia and Maine to Florida. It sometimes grows on upland soils where it is cultivated commercially. Cranberries (*Vaccinium macrocarpon, V. oxycoccus*, fig. 5.3) have long been known as a peatland crop. In some areas, bogs have been modified so they can be flooded for commercially harvesting cranberries. Cloudberry (*Rubus chamaemorus*) is a thornless member of the blackberry genus with fruits that are yellow or amber colored when mature. It is circumboreal in distribution and extends south only as far as New Brunswick, Maine, and New Hampshire in North America. It has been an important food for native Americans of northern regions. Other names that have been used for it are baked-apple berry, salmon berry, and yellow berry.

5.2. Highbush Blueberry (*Vaccinium corymbosum*)

5.3. Small Cranberry
(*Vaccinium oxycoccus*)

Sources of Fuel

Another age-old use of peatlands is as a source of fuel. The largest harvester of peat in the world today is Russia, where over 90 percent of the world's peat harvest occurs. Most of this is used for the generation of electricity. Russia also uses peat for home heating and in several other creative ways. For example, peat is mixed with cement to form peatcrete, lightweight building blocks used in construction. Peat has a long history of use for home heating in Ireland and northern Europe.

Peat Moss and Marl

The most important commercial use of peatlands in eastern North America is probably the production of peat moss for agricultural use. Sphagnum moss is usually the most abundant plant in peatlands. The leaf-like structures of this plant consist of very small living cells alternating with very large water storage cells. The latter give sphagnum moss a water-holding capacity of up to twenty-five times its dry weight. It has a greater water-holding capacity than absorbent cotton and it was used for surgical dressing during World War I. It is this absorbent quality that gives peat moss its great value as a mulching agent. Harvesting peat moss is a thriving business in some areas. Maine is one of the largest suppliers in the northeastern United States, but its annual production is only a fraction of that produced in New Brunswick.

Peatlands that develop in regions of limestone bedrock may have a water supply that is rich in calcium. Under certain conditions, the calcium will precipitate as calcium carbonate. This forms a deposit called marl when it is mixed with sand or clay. Marl can be harvested by draining or dredging, and some types are used in the manufacture of Portland cement. Clay marl can be used as an agricultural lime fertilizer for sour or acid soils.

Bog Iron

In some peatlands, the action of certain bacteria in the absence of oxygen causes the precipitation of iron compounds. These form a low-grade iron ore called limonite, or bog iron. The mining and processing of bog iron was a flourishing industry in the New Jersey pine barrens from the 1760s to the mid 1800s. One of the largest operations was the Batsto Iron Works founded

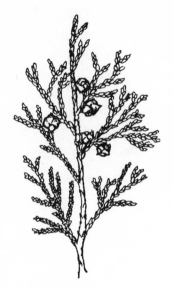

5.4. Atlantic White Cedar
(*Chamaecyparis thyoides*)

in 1766. During the American Revolution, it provided cannon balls for General Washington's army and was almost captured by the British in 1778. The bog iron industry reached its peak during the War of 1812, then declined when higher-grade iron ores near substantial coal deposits were discovered in western Pennsylvania. The Batsto Iron Works closed in 1848, but today tourists can view exhibits on the bog iron industry at the Batsto Village State Historical Site west of Atlantic City.

The Geography of Peatlands

Canada contains within its borders about 40 percent of the world's peatlands. With more than 600,000 square miles (1,627,000 sq. km), it leads the world in this resource, which is referred to in Canada as muskeg. The Great Bear–Great Slave Lake Region and the Hudson Bay lowlands make up part of this great expanse. Russia has about 36 percent of the world's peatlands with 579,000 square miles (1,505,400 sq. km). The United States is a distant third with only about 7 percent and 116,000 square miles (301,600 sq. km) of peatlands. Most of the American peat is found in Alaska and Northern Minnesota. Low rates of decomposition because of low average annual temperatures and poor drainage result in vast areas of arctic tundra peatlands. In the lower forty-eight states, the single greatest expanse of peatlands is in the four counties north of the Red Lakes region of Minnesota.

In eastern North America, peatlands include the pocosins, discussed in chapter 1, inland bogs, which will be discussed in a later section, and Atlantic white cedar bogs. The latter occur in a band within 125 miles (200 km) of the coast from New Brunswick and Maine to Long Island, New York, and along the coastal plain to Mississippi. They are most commonly called cedar swamps or cedar bogs but local names include spungs or spongs in the New Jersey pine barrens, juniper lights in the Great Dismal Swamp of Virginia, and juniper bogs throughout the south. Regardless of their local designa-

tion, a species they all have in common is Atlantic white cedar (*Chamaecyparis thyoides*, fig. 5.4).

Atlantic white cedar is a tree that
may grow to 83 feet (25 m) in height
with a trunk 4 feet (130 cm) in diameter. The leaves are scale-like, overlapping, lying flat against the twig. Male
and female cones are produced on the
same tree with small winged seeds released September to October and dispersed by wind. The wood is decay
resistant and insect repellent, which
in the past made it very useful for
telephone poles, roof shingles, siding,
fence posts, and railroad ties. It is of
little economic importance today because excessive lumbering has almost
eliminated it from the market. Other

5.5. Sweet Pepper-Bush (*Clethra alnifolia*)

names that have been used for Atlantic white cedar are coastal cedar, white
cedar, false cypress, swamp cedar, southern white cedar, and juniper.

Coastal plain peatlands may include bogs with pure stands of Atlantic
white cedar that dominate the canopy. In these, sweet bay (*Magnolia virginiana*, fig. 4.19), red maple (*Acer rubrum*, fig. 4.20) and black gum (*Nyssa sylvatica*, fig. 4.25) may be understory trees. Sometimes all the latter species
share the canopy in mixed bog-forest stands. Farther north, species that
may be associated with Atlantic white cedar are white pine (*Pinus strobus*)
and hemlock (*Tsuga canadensis*). In southern states, bald cypress (*Taxodium
distichum*, fig. 4.27) and water tupelo (*Nyssa aquatica*, fig. 4.28) may be
associates.

Atlantic white cedar bogs that have a relatively open canopy usually
have a well-developed shrub zone. Members of the heath family (*Ericaceae*)
are the most abundant shrubs. There is some variation in floristic composition from north to south, but some species have a wide geographic range.
These include highbush blueberry, sweet pepper-bush (*Clethra alnifolia*, fig.
5.5), and inkberry (*Ilex glabra*, fig. 1.4). Shrubs that have a more northern
distribution are large and small cranberries, leatherleaf (*Chamaedaphne caly-*

5.6. Leatherleaf
(*Chamaedaphne calyculata*)

5.7. Bog Rosemary (*Andromeda glaucophylla*)

culata, fig. 5.6), and bog rosemary (*Andromeda glaucophylla*, fig. 5.7). Species not normally seen in northern bogs are wax-myrtle (*Myrica cerifera*, fig. 4.31) and fetterbush (*Lyonia lucida*, fig. 1.2).

On the herbaceous level, sphagnum moss is common throughout coastal white cedar bogs. In areas where the canopy is open enough for sunlight to reach the surface, bog meadow species flourish. Among these are the insectivorous species: sundews, pitcher plants, bladderworts, and butterworts. These will be discussed in a later section. Common in peatlands of the coastal plains are several ferns including Virginia chain-fern (*Woodwardia virginica*, fig. 1.6), cinnamon fern (*Osmunda cinnamomea*, fig. 2.8), and royal fern (*O. regalis*, fig. 5.8).

Bogs of the Northeast

How They Came to Be

Bogs are common features of the landscape in all parts of the northern hemisphere that were covered by ice in the last fifteen thousand years. They are wetlands that support a specialized community of herbs and shrubs growing on a water-saturated substrate composed of peat. As the glaciers retreated northward, large chunks of ice were buried by rocks and soil carried in the ice. When the ice melted, depressions called kettles were formed, and many filled with water to become small lakes and ponds. Bogs developed in many of these bodies of water.

The bog probably started forming as a fringe of sphagnum moss and other herbaceous plants along the margins of the water-filled depression. Under cool, moist conditions that were ideal for growth, the sphag-

5.8. Royal Fern (*Osmunda regalis*)

num fringe expanded and began to creep out over the water in floating mats. Each year new growth extended the mat farther into the pond and pushed last year's plants deeper into the water. Year after year the floating sphagnum mat closed in on the center of the pond and the basin progressively filled as new growth forced last year's growth downward. After thousands of years, the open water disappeared entirely yielding the quaking bog we see today.

Although not as numerous as those in glaciated areas, there are bogs in southern regions. They have developed in basins formed by forces other than glaciation. For example, Cranberry Glade Botanical Area in southern West Virginia is a group of high-elevation bogs in a basin caused by the erosion of a soft layer of rock overlying a tilted harder layer. However, the method of growth of the vegetation in these and other southern bogs was probably similar to that of northern ones.

Peat deposits sometimes accumulate in places other than water-holding depressions. In the Pocosins of North Carolina, peat has accumulated on sloping terrain. These and Cranberry Glades were discussed in greater detail in chapter 1.

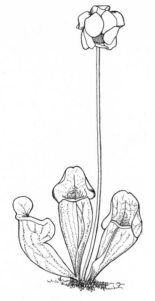

5.9. Pitcher-Plant (*Sarracenia purpurea*)

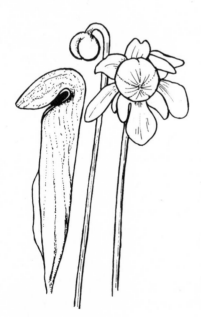

5.10. Hooded Pitcher-Plant (*Sarracenia minor*)

The Bog Environment

The water level is very near the surface in most bogs and walking on one is like walking on a waterbed. Each step causes a wave of motion that is transmitted to the surrounding plants causing them to vibrate, hence the name quaking bog. In order to maintain dry feet, the intrepid naturalist who ventures onto a bog will usually need at least calf-high boots. Even so, soft spots where a step may plunge the leg into wet peat above the knee will convey the true meaning of being "bogged down." Bogs that have floating mats can be hazardous because falling through one can be a very wet experience. Most bogs are safe to walk on, but some national and state preserves have boardwalks from which the plants can be observed without damaging the bog ecosystem.

A characteristic feature of bogs is a thick carpet of vegetation consisting of several layers of sphagnum moss. Metabolic activity of the moss causes the accumulation of acids making the bog a very acidic environment. Since the water level is near the surface, the peat is saturated and essentially free of oxygen. These acidic, oxygen-free conditions prevent the growth of bacteria and fungi that would ordinarily break down the accumulated organic matter. The result is that bogs have great preservative qualities, and the plants that make up the peat decompose very slowly.

Special Groups

Insectivorous plants. Among the specialized groups of plants that grow in bogs are those that trap insects. These plants carry on photosynthesis like other green plants, but supplement their nutrition by capturing and digesting small insects. Although it has not been proved conclusively, this practice may have evolved in response to mineral deficiencies in the bog substrate. In mineral soils, nitrates and other nutrients are present from the breakdown of rocks and the decomposition of plants. In the bog habitat, where there is no mineral soil and decomposition is very slow, a chronic shortage of mineral nutrients exists. The following are insectivorous plants common in many bogs.

Pitcher plant has hollow pitcher-shaped leaves that collect rain water. Insects fall into the water and are digested by enzymes produced in the leaf. The most common species in northern bogs is *Sarracenia purpurea*, (fig. 5.9). Species in southern bogs are trumpet pitcher plant (*S. flava*) and hooded pitcher plant (*S. minor*, fig. 5.10).

Sundew has small roundish leaves covered with hairs, each with a

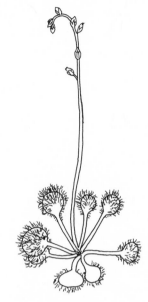

5.11. Round-Leaved Sundew (*Drosera rotundifolia*)

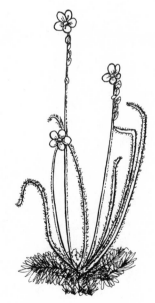

5.12. Thread-Leaved Sundew (*Drosera filiformis*)

drop of a sticky substance at its tip. When a gnat or other small insect becomes trapped on the sticky hairs, the leaf slowly folds over it like a closing hand. It reopens, perhaps days later, after the insect is digested. In addition to round-leaved sundew (*Drosera rotundifolia*, fig. 5.11) and spatula-leaved sundew (*D. intermedia*), which are common in northern and western bogs, southern bogs may also contain thread-leaved sundew (*D. filiformis*, fig. 5.12).

Butterwort has a rosette of basal leaves with a leafless flower stalk rising from the center. The leaves have a sticky coating in which small insects are trapped and digested. Common butterwort (*Pinguicula vulgaris*, fig. 5.13) with blue flowers occurs in northern and western bogs and yellow butterwort (*P. lutea*) is a southern bog species.

Bladderwort is found in bogs with open pools of water in their centers and also in swamps. It has finely dissected leaves with many tiny bladders (fig. 5.14). Each bladder has an opening with a ring of hairs at one end. When a small aquatic organism, such as a water flea, touches one of the hairs, the bladder inflates suddenly, sucking in water along with the hapless victim. Once the insect is inside, the opening closes and the organism is digested.

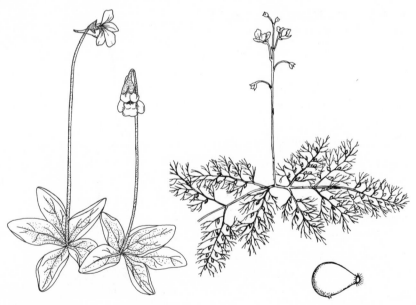

5.13. Common Butterwort (*Pinguicula vulgaris*)

5.14. Bladderwort (*Utricularia* sp.)

Orchids. Several native orchids prefer the environmental conditions found in bogs. Orchids evolved in the tropics, and the greatest number of species grow there today, typically on the trunks of trees. There is some similarity with regard to nutrient availability between this habitat and the one occupied by bog orchids. Common orchids found in bogs are grass-pink (*Calopogon tuberosus*, fig. 5.15), rose pogonia (*Pogonia ophioglossoides*, fig. 5.16), moccasin flower (*Cypripedium acaule*, fig. 5.17), and white fringed orchid (*Habenaria blephariglottis*, fig. 5.18).

In order to see these plants in bloom, the naturalist should visit the bog in late June or early July to observe grass-pink, rose pogonia, and moccasin flower; and in July or August for the white fringed orchid. Grass-pink has flowers two inches wide, pink to deep rose, with a crest of yellow hairs. Rose pogonia has rose-pink flowers with yellow hairs on the central petal. Moccasin flower has a large, pouch-like central petal that is pink with deeply colored veins. White fringed orchid has several flowers with a long spur and a fringed upper petal at the tip of the stem.

Shrubs. The most abundant shrubs growing in bogs are low-growing

5.15. Grass-Pink (*Calopogon tuberosus*); 5.16. Rose Pogonia (*Pogonia ophioglossoides*); 5.17. Moccasin Flower (*Cypripedium acule*); 5.18. White Fringed Orchid (*Habenaria blephariglottis*)

5.19. Bog Laurel (*Kalmia polifolia*)

5.20. Labrador Tea (*Ledum groenlandicum*)

evergreen members of the heath family. Evergreen plants are more efficient at using the scanty mineral nutrients in bogs because they do not shed all their leaves each year as do deciduous plants. Even though bog shrubs are usually growing with their roots in water, they have leaf characteristics of xerophytes. These include waxy coatings; rolled-under margins; and scaly, fuzzy, or woolly undercoatings. Recent investigations indicate that these water-conserving features may be a response to the low nutrient level of the bog habitat.

The evergreen bog shrubs are leatherleaf (fig. 5.6), bog rosemary (fig. 5.7), bog laurel (*Kalmia polifolia*, fig. 5.19), Labrador tea (*Ledum groen-landicum*, fig. 5.20), large cranberry, and small cranberry. All of these except leatherleaf and large cranberry have leaves with margins rolled under. Leatherleaf has leaves with an undercoating of brown scales; bog rosemary has blue-green leaves with white undersides; Labrador tea leaves have dense woolly undersides; and bog laurel has dark green, shiny leaves with white undercoatings. The leaves of the cranberries are very small with white undersides. Those of small cranberry are smaller and are usually rolled under at the margins. Flower stalks of large cranberry arise from the

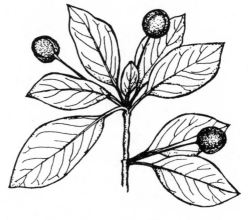

5.22. Bog Holly (*Nemopanthus mucronatus*)

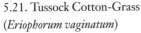

5.21. Tussock Cotton-Grass
(*Eriophorum vaginatum*)

axils of reduced leaves toward the base of the stem while those of small cranberry are attached at the tips of the stems.

Life History of a Bog

In the beginning of bog development, as the floating sphagnum mat expands, it supports sedges and other herbaceous plants like the orchids and insectivorous plants. This is a bog meadow, the earliest stage in the life of a bog. Sedges are monocotyledons that usually have three-sided stems with rows of grass-like leaves on each side. Several species of sedges normally grow in bogs, but one of the most conspicuous is cotton grass. Tussock cotton-grass (*Eriophorum vaginatum*, fig. 5.21) has a tuft of white cottony bristles at the tip of a stem with greatly reduced leaves. As they wave in the breeze, these snowy white tufts give the bog its autumnal aspect.

The seeds of shrubs usually germinate among the herbs, or, more commonly, closer to the shoreward margin of the bog. Over a long period of time, as the basin fills with peat, the shrubs will advance toward the center of the bog and eventually cover the entire surface. As the shrubs increase in number, the shade beneath them becomes more dense and the herbaceous plants decrease or disappear. Large deciduous shrubs that may be part of the plant cover are highbush blueberry and bog holly (*Nemopanthus mucronatus*, fig. 5.22).

5.23. Poison Sumac (*Toxicodendron vernix*)

Along with the shrubs, the seedlings of several pioneer tree species may be observed. These include larch (*Larix laricina*), black spruce (*Picea mariana*), red maple, and poison sumac (*Toxicodendron vernix*, fig. 5.23). After many years, a bog forest will develop that may persist as long as the water level remains the same. In coastal New England bogs and southward along the coastal plain, Atlantic white cedar is often a prominent tree of the bog forest.

A bog is a harsh environment for pioneer trees, and they grow very slowly. A tree only a few feet tall may be a hundred or more years old. As a consequence of the water-logged substrate, root systems are shallow and trees are easily toppled by strong winds. A series of wet years can raise the water level enough to kill most of the trees. It is thus not unusual to see a number of dead trees in a bog. Even after a bog forest has developed, a high water level for several years may cause a reversion to the shrub or bog meadow stage.

As the result of erosion or some other natural or man-made force, drainage patterns may change and cause the water level in a bog to drop. When this happens, the peat will dry and may be harvested as peat moss. When bog succession has progressed to the forest stage, the surface peat has oxidized and decomposed to a fine black consistency. In many places in eastern North America, the trees have been removed and drainage ditches dug to create muck farms. Muck farming is economically important in some regions for the production of vegetable crops such as lettuce, onions, and celery.

Nature's Historical Museum

From the earliest of times, people have considered bogs to be strange and forbidding places. One early writer even referred to them as "trysting places for witches." They are indeed unique habitats, and their unfamiliarity has probably been responsible for these reactions of fear and foreboding. A contributing factor has been their preservative qualities. Like natural mausoleums, bogs have yielded the preserved bodies of long-extinct Ice Age mammals such as mastodons and mammoths.

In Europe, where peat is often dug as a fuel, preserved two-thousand-year-old bodies of humans have been uncovered that seem to have been buried as a ritual or as a form of execution. Tollund Man is an example of a body uncovered in Denmark in 1950. The face was so well preserved that the peat harvesters thought they had stumbled on a recent murder victim. The body was buried in five feet of peat and wore only a peaked leather cap with a chin strap, a leather belt, and a noose around his neck. The cause of death had been strangulation or hanging. Analysis of his stomach contents indicated that his last meal contained no summer or fall fruits, suggesting he died in winter. P. V. Glob, who wrote a book entitled *The Bog People*, thinks Tollund man may have been sacrificed at a midwinter ritual in supplication to the goddess of fertility for successful crops. The head of Tollund man has been preserved and is on display near where he was found at the Silkeborg Museum in central Denmark. Over thirty bodies of bog people have been recovered from peat bogs in Denmark.

The preservative nature of bogs has been very useful to scientists studying the history of vegetation. During the thousands of years it took for the bog basin to fill with peat, the pollen from local plants became part of the yearly accumulation. Wind-pollinated plants produce an abundance of pollen that is blown about. Some is deposited in bogs. Using a special kind of sampling device, biologists have been able to take samples of peat from all layers of the bog. Fortunately, each genus of tree has pollen with distinctive characteristics. By identifying the pollen in any layer, the botanist can identify the trees that were growing in the vicinity of the bog when the peat at that layer was formed. After the trees have been identified, conclusions can be drawn about the climatic conditions necessary for their growth. Studies of this nature have resulted in detailed knowledge of the forests and climates that have existed in North America since the retreat of the last glacier.

6

Plants of Special Interest

Early humans were much more familiar with the plants around them than the average person today. Wild plants made up a large portion of their daily food. Medicines administered by shamans, witch doctors, or medicine men came mainly from plants. Religious rituals were often accompanied by the consumption of hallucinogenic plants. Only the medicine man needed to know the healing and ceremonial plants, but the knowledge of what could and could not be eaten was, of necessity, more widespread. Few modern Americans would survive if their lives depended on finding wild food and medicinal plants. This chapter will provide brief discussions of some poisonous, medicinal, hallucinogenic, and edible wild plants that may be seen on a field trip into a freshwater wetland.

Poisonous Plants

Poisonous plants are those containing substances that have harmful affects on the human body if they come into contact with the skin or are eaten. Considering that there are at least 300,000 species of plants, the percentage of known poisonous ones is relatively small. Nevertheless, each year hundreds of cases of poisoning are reported to Poison Control Centers in the United States and Canada. Most involve children who nibble on poisonous houseplants or sample the plant fare in their back yards. Poisoning in adults most often results from misidentifying a poisonous plant and using it for food or an herbal remedy.

There are no reliable physical characteristics that can be used to distinguish poisonous from nonpoisonous plants. Some writers have suggested that plants with red or white berries, milky sap, or an unpleasant odor should be avoided as potentially poisonous. To be sure, there are poisonous plants with these features, but other toxic plants have blue berries; orange, red, or colorless sap; or the pleasant odor of parsnips. To make matters worse, there are harmless and even edible wild plants with these same characteristics.

A belief held by many is that if one can observe other animals eating a plant it is safe for human consumption. This can be a fatal assumption. Cattle and horses have been fatally poisoned by eating the yew plant (*Taxus* spp.) several species of which are widely planted ornamental shrubs. These plants are also highly toxic to humans. Although many species of birds eat the berries of poison ivy without apparent harm, it would be very dangerous for humans to consume even one. Therefore, it is not a safe practice to rely on generalities for the identification of poisonous plants. If they are being collected for human use, either as food or for home remedies, any plant that is unknown to the collector should be left where it stands.

Plant Poisons

Functions of plant poisons. Biologists who study plant evolution are interested in determining the origin of the physical and chemical traits of plants. Most characteristics have evolved in response to environmental conditions and contribute to the survival of the species. There is still a lot to be learned about why plants produce poisonous substances, but there are several possibilities. One is that they are waste products of metabolism. Since plants do not have excretory systems, wastes cannot be eliminated, as in animals, but must be stored in some part of the plant. Another possibility is that the poisonous substances are compounds that are essential in the normal metabolism and maintenance of the plant and their toxicity to humans and other animals is a coincidence. A third possibility is that poisonous substances have evolved as defense mechanisms against their greatest natural enemies, plant-eating insects.

It is not clear whether urushiol, the poisonous substance of poison sumac (*Toxicodendron vernix*) is a waste product, an essential metabolic compound, a defensive insect repellent or none of these. It is an accident of nature that urushiol is toxic to most humans.

Types of plant poisons. An important and widespread group of plant poisons are called alkaloids. These are compounds that contain nitrogen and react chemically as bases rather than acids. They are almost always bitter tasting and may be present in up to 40 percent of all plant families. Most alkaloids produce a strong reaction on the nervous system when ingested by animals, including humans. This action makes some of them highly toxic, but some are also very important medicinally. The names of alkaloids always end in *ine* or *in* and they are often named for their plant source: veratin for *Veratrum viride*, false hellebore, and lobeline for *Lobelia cardinalis*, cardinal flower.

Other types of poisonous substances in plants are oxalic acid and oxalates, phenols, polypeptides, resins, and poisons accumulated from minerals in the soil. These compounds include the poisons in such plants as poison sumac, water hemlock (*Cicuta maculata*), and poisonous mushrooms. For more information on these and other plant poisons, see Kinsbury (1964), Kinghorn (1979), and Harden and Arena (1974).

Types of Reactions to Plant Poisons

Skin Irritations

The plants that most commonly cause skin irritations or dermatitis in North America are members of the genus *Toxicodendron*. The only species commonly found in freshwater wetlands is poison sumac (*T. vernix*, fig. 5.23). It is a shrub or small tree of swamps and marshes throughout most of eastern North America. All members of the genus, which includes poison ivy and poison oak, contain a substance known as urushiol, an oily resin, to which most people are allergic. This toxin is a colorless or milky fluid within special canals in all parts of the plant except the pollen.

For individuals sensitive to the toxin it might be helpful to review several facts. (1) You cannot get a reaction from simply touching a leaf or stem. The plant part must be bruised or broken so that the special canals are ruptured and the toxin comes in contact with the skin. (2) Dead leaves and stems will cause a reaction as readily as green ones. (3) The plant should never be burned because the vaporized toxin and particles in the smoke may affect the eyes, nose, and lungs.

A traditional remedy for exposure to poison sumac is to wash with strong soap as soon as possible after contact. According to the *American Medical Association Handbook of Poisonous and Injurious Plants*, this is not a good idea. It takes about ten minutes for the toxin to penetrate the skin. If a strong soap is used, the natural body oils will be removed and any remaining toxin may penetrate even faster. Since urushiol is not soluble in water, the recommended treatment is to wash with plain running water without soap.

A common misconception is that the fluid from the blisters of the dermatitis will spread the rash. When the toxin penetrates the skin, it combines chemically with deeper skin tissues. All of the toxin interacts with the cells and thus the greater the exposure, the more severe the rash. Since all of the toxin undergoes an irreversible chemical change, there is none left in the fluid of the blisters, so the rash cannot spread when the blisters burst.

There are commercial lotions that claim to protect the user from the urushiol toxin. For those who are allergic, though, the best practice is to learn to recognize the plants and stay away from them.

Internal Poisoning

Fatal poisoning of adults is usually the result of mistaking a poisonous plant for an edible one. The ingestion of the water hemlock, poison hemlock (*Conium maculatum*, fig. 6.1), and poisonous mushrooms have been responsible for numerous deaths of both children and adults. Water hemlock (fig. 6.2) and bulb-bearing water hemlock (*C. bulbifera*) may be the most deadly native herbaceous plants in North America. They are both perennial herbs that grow in swamps, marshes, pond margins, and roadside ditches. Bulb-bearing water hemlock is more common in the northeastern and northwestern parts of North America but water hem-

6.1. Poison Hemlock (*Conium maculatum*)

6.2. Water Hemlock (*Cicuta maculata*)

lock is common throughout. The poisonous substance in these plants, a yellow oily liquid, is a complex alcohol that is concentrated in the roots. The root has a rather pleasant smell of parsnips and it has been mistaken for the root of wild parsnip (*Pastinaca sativa*). A single bite is enough to kill an adult human, and a bite the size of a walnut is enough to kill a cow.

Poison hemlock is not a native plant but was introduced from Europe. It contains conine and several other poisonous alkaloids in all parts of the plant but concentrated in the leaves and seeds. It grows along roadsides and in fields, pastures, damp meadows, and wetland borders throughout most of the United States and southern Canada. Poisoning in humans is usually the result of mistaking the young leaves for parsley or the seeds for anise or dill. Poison hemlock is also found in Europe and Asia, and it was used in earlier cultures to execute criminals. It causes death by paralysis of the respiratory system. The philosopher Socrates in ancient Greece was sentenced to drink a cup of poison hemlock tea for his execution. Water hemlock and poison hemlock are not related to the hemlock tree, which is nonpoisonous.

Mushroom Poisoning

Of all the members of the plant and fungi kingdoms, the mushrooms are the most notorious as poisoners. This reputation is justified for some species, but as with green plants, only a small percentage of the total number of mushroom species are known to be toxic. There are several thousand species of mushrooms in the United States. It is estimated that about one hundred species bring about harmful reactions when ingested, and no more than ten are deadly poisonous. Although all species have not been tested for toxicity, it is believed that these estimates are not likely to change. In folklore, poisonous mushrooms are sometimes referred to as toadstools. The word is derived from a German word that means death's stool. It is not a scientific term.

As with poisonous green plants, there are no rules that a novice can follow for identifying poisonous mushrooms. The differences between edible and poisonous species are often so slight that it requires a trained expert to tell them apart. In addition, a single cluster of mushrooms may include edible and poisonous species growing side by side. This does not seem to deter persistent mushroom gatherers. Consequently, there are hundreds of cases each year of mushroom poisoning and two or three fatalities. The wild mushroom enthusiast should keep in mind a bit of pithy folk wisdom:

> There are old mushroom hunters
> There are bold mushroom hunters
> But there are no old, bold mushroom hunters.

There is great variation in the ways humans react to mushroom poisoning. No single set of symptoms can be associated with all cases. Species that cause a reaction in some people are eaten by others without problems. The degree of toxicity of a species may depend on the season it is collected, the geographic area in which it grows, or the health of the consumer. Unlike many of the poisonous green plants, the poisonous mushrooms apparently do not have a bad odor or a bitter taste. A survivor of poisoning by one of the most deadly mushrooms reported that it was delicious.

Emetic russula (*Russula emetica*, fig. 6.3) is a wetland species that is usually considered to be poisonous. It is a medium-sized mushroom with a bright red to deep pink cap. The cap is sticky and may be 4 inches (10 cm) across with a stalk to 4 inches high. The stalk widens toward the base and there is no ring. It can be observed from August to September, commonly on sphagnum bogs but also in woodlands. The toxic substances in this mush-

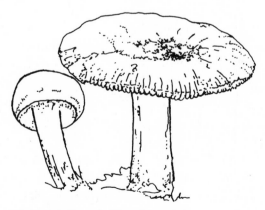

6.3. Emetic Russula (*Russula emetica*)

room are not fully understood and they do not affect everyone in the same way. The symptoms of poisoning appear in less than an hour and include nausea, stomach cramps, and vomiting. Poisoning is usually not fatal, and it is even collected as an edible mushroom by some. But because of the lack of knowledge about its toxins, this mushroom should be avoided. The best advice that can be offered to any amateur mushroom hunter is to restrict mushroom consumption to those that are available from the canned goods or produce section of the supermarket.

Poisonous Plants in the Field

Algae

None of the freshwater green algae are known to be poisonous but the same cannot be said for the blue-green algae. In warm, quiet waters very rapid growth of the blue-greens can take place in what is known as an algal bloom. Evidence of this can be observed as slimy masses of algae floating on the water surface. Sometimes in low-water conditions this occurs in reservoirs, giving the water an unpleasant odor. There are many cases of livestock and a few of human poisonings, and sometimes death, from toxic blue-green algae in the water supply.

Ferns

Most ferns are harmless, but there are a few species that are known to be poisonous to livestock and humans. Sensitive fern (*Onoclea sensibilis*, fig. 6.4) grows in marshy areas, wet meadows, and damp forest margins throughout eastern North America. The common name refers to its sensitivity to cold. It is one of the first plants to wither in autumn when night temperatures drop to the low thirties. The substance in the fern that is toxic is unknown, but in feeding experiments this fern has been highly toxic to horses.

6.4. Sensitive Fern (*Onoclea sensibilis*)

Horsetails

Marsh horsetail (*Equisetum palustre*) has been shown to be toxic to cattle and horses. It is most frequently found in wet or moist habitats in the northern United States and southern Canada. It contains an enzyme that destroys vitamin B1. It is not likely that humans could be poisoned by horsetails, but some herbal remedies prescribe a tea made by steeping the plant in boiling water. Such use of these plants could lead to a deficiency of vitamin B1.

Gymnosperms or Conifers

Yew. This is an evergreen shrub or small tree. It has flat needles that usually grow in two rows along the twigs. The most common wild species in the northern United States and southeastern Canada is American yew or ground hemlock (*Taxus canadensis*, fig. 6.5). It grows in moist to wet woods and around the margins of bogs. Reproductive structures of all the yews are fleshy, bright red berries each surrounding a single seed. All parts of these plants are poisonous except the reddish tissue of the berry; the seed itself is also toxic.

Herbaceous Flowering Plants

Poison hemlock is a biennial that blooms from June to August with many tiny white flowers in flat-topped clusters at the ends of stems and branches. The leaves are alternate, finely dissected, with stalks that are enlarged at the base and clasp the stem. The stem is hollow, usually with purple spots, reaching 6 feet (1.8 m) in height. It is somewhat similar to wild carrot (*Daucus carota*) but has a smooth stem while wild carrot has a hairy stem. Poison hemlock is slightly less poisonous than water hemlock but is still deadly.

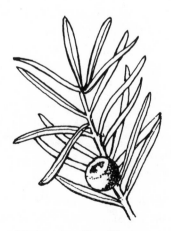

6.5. American Yew (*Taxus canadensis*)

  Cardinal flower (fig. 6.6) is a perennial that blooms from July to September. The flowers are scarlet and two-lipped with three

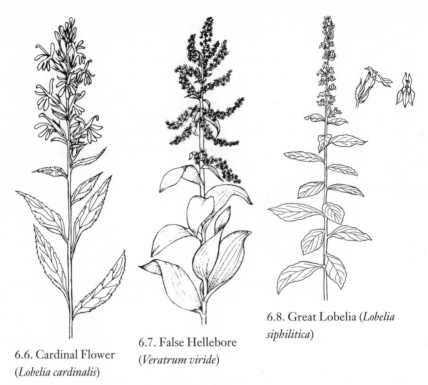

6.8. Great Lobelia (*Lobelia siphilitica*)

6.7. False Hellebore (*Veratrum viride*)

6.6. Cardinal Flower (*Lobelia cardinalis*)

prominent lobes in the lower lip. The flowers are in an elongated cluster at the tip of the stem. The leaves are alternate, lance shaped, on an unbranched stem that may be 5 feet (1.5 m) high. Cardinal flower grows in wooded swamps on pond and stream margins and in wet woods throughout eastern, central, and southern North America. Plants of this genus have several toxic alkaloids. They have been used in herbal remedies for a variety of disorders, and a number of human deaths have resulted from overdoses. Death of livestock is rare, but feeding experiments have shown these plants to be poisonous to sheep and other animals.

False hellebore (fig. 6.7) is a perennial that blooms in June and July with many green, six-petaled flowers in a branched cluster at the tip of the stem. The leaves are alternate, to 12 inches (30 cm) long, with prominent parallel veins. The straight, unbranched stem may be $6^1/_2$ feet (2 m) high. False hellebore grows in swamps and wet woods from New Brunswick and Quebec to Alaska, south to Georgia in the east, and to Oregon in the west. All parts of the plant contain complex poisonous alkaloids. Poisoning in humans is most often the result of an overdose or misuse of an herbal remedy.

Early American settlers used a solution made by boiling the roots in water as an external application for herpes infections and to kill head lice.

Great lobelia (*Lobelia siphilitica*, fig. 6.8) is a perennial that blooms from July to September with blue two-lipped flowers, the upper lip having two small lobes, the lower lip three larger lobes. The leaves are alternate and lance shaped with the widest point near the tip. The stem is unbranched and may be up to 5 feet (1.5 m) high. Great lobelia grows in swamps, along stream and pond margins, and in open wet woods throughout most of the eastern United States and southeastern Canada. For the poisonous qualities of great lobelia, see the related species, cardinal flower.

Water hemlock (fig. 6.2) is a perennial that blooms from June to September with tiny white flowers in flat-topped clusters at the tip of stems and branches. The leaves are alternate, two or three times pinnately compound, with leaflets having pointed teeth and veins ending in the notches between teeth. The stem is thick, hollow, freely branched, often with purple splotches or lines, and may grow to a height of $\frac{1}{2}$ foot (2 m).

Toxic Plant Ingestion: What to Do

Although children are most often the victims of poisoning by the ingestion of toxic plants, adults are also sometimes poisoned. The best treatment is to avoid poisoning altogether. For children, keep toxic plants out of reach. For adults, become familiar with the poisonous plants in your area and *never* consume a wild plant or mushroom unless you are *sure* of its identity. Never take a herbal remedy unless you are sure of the identity of the plants used in its construction. When a suspected poisonous plant or mushroom has been ingested, do not waste time trying to identify the specimen. GO IMMEDIATELY TO THE EMERGENCY ROOM OR CALL A PHYSICIAN, OR THE LOCAL POISON CONTROL CENTER. If at all possible, have a sample of the suspected poisonous specimen available.

If an emergency room or a physician cannot be reached, the best practice in most cases is to induce vomiting. If a finger or blunt instrument in the back of the throat does not succeed, syrup of ipecac can be used. It contains several alkaloids that cause vomiting, and it is available as a nonprescription drug. It should be taken as soon as possible after ingestion of the suspected poisonous plant or within two hours. Ipecac should not be administered to a person who has lost the gag reflex, is not fully conscious, or shows signs of convulsions. The stomach contents should be saved, especially if a specimen of the suspected poisonous plant or mushroom is not available.

The recommended dose of ipecac for adults is two tablespoons (30 ml) and for children over one year old, one tablespoon (15 ml). These doses may vary with the individual and may not be appropriate for everyone. Ipecac should be taken with a glass of water or some liquid other than milk. For children under one year of age, ipecac should be administered only under the direction of a physician. Most people are within at least telephone distance of qualified medical assistance. The instructions presented here are under no circumstances to be followed rather than calling for medical assistance.

Hallucinogenic Plants

Definitions and Beginnings

Hallucinogenic plants contain compounds that act on the central nervous system. They bring about changes in mood and distort the ways in which time, space, color, and sound are perceived. These departures from reality are called hallucinations. Most of the known hallucinogenic plants are in the dicotyledon group of the angiosperms or flowering plants. There are none in the gymnosperms, ferns and fern allies, mosses and liverworts, or algae. There are several species of fungi, however, that contain hallucinogenic substances.

Humans have known about and used hallucinogenic plants for thousands of years. It is interesting to speculate on how early man may have learned to distinguish between poisonous, medicinal, and hallucinogenic plants. In the learning process, no doubt, many became ill and probably many died. Then as now some individuals were more perceptive than others, and these became the shamans or medicine men. In their earliest uses, the hallucinogenic substances served as a means of communicating with the spirit world for guidance in times of crisis. Every known hallucinogenic plant has a history of such use in early cultures.

Consulting with the spirit world was a very serious matter in ancient cultures as well as in some modern primitive cultures. It was undertaken solemnly and often with elaborate ceremony. In his visions, the medicine man would look for answers to religious, medicinal, social, or military problems. Some writers have suggested that the very concept of God originated in these visions. During these times, the use of hallucinogenic substance was limited mainly to the medicine men and never a practice among the common people. Only in relatively recent times have these substances been subjected to widespread recreational use and often abuse.

Basing the decision solely on the compounds they contain, it is sometimes difficult to distinguish between poisonous, medicinal, and hallucinogenic plants. In this chapter, plants are arbitrarily listed in one or the other of these categories only for purposes of discussion. The poisonous compounds in some plants are important medicines when taken in controlled doses. Some hallucinogenic substances in plants are important medicines when taken at one amount but are deadly poisonous when taken in larger doses. The hallucinogenic plants are not described here to suggest experimentation. One needs only to read the daily newspaper for accounts of fatal overdoses with hallucinogenic substances.

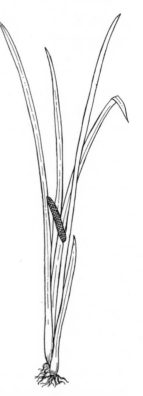

6.9. Sweet Flag (*Acorus calamus*)

Sweet flag (*Acorus calamus*, fig. 6.9) is a perennial that blooms from May to August with many tiny, greenish-yellow flowers in a cone-shaped structure projecting from the side of a leaf-like stem. The leaves all arise from the base of the plant and are long, narrow, ribbon-like, to 4 feet (1.2 m) long. The flower-bearing stems are leaf-like and may be 32 inches (80 cm) in height. Sweet flag is found in marshes, open swamps, around the margins of ponds, and in roadside ditches from Nova Scotia and Quebec to Washington State, and south to Florida, Texas, and Colorado.

Sweet flag has been used both as a wild food plant and a medicinal herb. The young inner shoots have been recommended for salads. The peeled rootstock can be boiled and candied for a confection with a distinctive flavor. In herbal medicine, the powdered root has been used as a dentifrice. A solution made from boiling the root in water has been used for upset stomach, intestinal gas, a treatment for coughs and for relief of cold symptoms. Some American Indian tribes drank the solution to bring about abortions. The Cree Indians chewed the root as a strong stimulant and hallucinogen.

An oil extracted from the rootstock contains asarone, which is an hallucinogen. The Food and Drug Administration prohibits the sale of this oil. In a laboratory study, this substance caused tumors in rats. Until more information is available, it is prudent to avoid this plant as food or as an herbal remedy.

Medicinal Plants

Their Importance

The shamans and medicine men of primitive cultures were probably the first professional men. Since most of the medicines they dispensed came from plants, they, of necessity, were botanists. This strong bond with plants by those that practice the healing arts has been a characteristic of human societies from prehistoric times. It continued into modern times until near the end of the 1800s. Even at that date, many medical doctors were botanists and most professional botanists were physicians. In 1900 about 80 percent of the drugs prescribed by physicians came directly from plants. The growth of organic chemistry, beginning at about this time, initiated an era of synthetic medicines. Although the development of synthetic medicines has continued into present times, 35 to 40 percent of all prescribed drugs are still either natural plant compounds or plant compounds in combination with synthetic substances.

Plants are still as important in the practice of medicine as they were to the shamans and medicine men. Today they serve the medical profession in at least three ways. One way is that almost 25 percent of the drugs prescribed by modern physicians come directly from plants. Secondly, some plant compounds are used as essential components in the manufacture of medicinal drugs. Finally, natural plant drugs may serve as models for the synthesis of identical or similar drugs.

Herbal Medicine

In the early days of colonization in North America, physicians and hospitals were few or nonexistent. The settlers had no choice but to rely on herbal medicine for treating illness and injuries. They had brought with them a rich heritage of herbal remedies from Europe, and they soon added to these by including treatments learned from American Indians. The result is that there are folk remedies for almost every ailment experienced by humans.

A complete description of all the plants, and the uses that have been made of each in herbal medicine, would require a large book. Very few of these remedies have been subjected to controlled testing to verify their effectiveness. The use of some plants can be traced to the ill-conceived doctrine of signatures, which held that the shape of a leaf, root, or seed determined its use in healing. Other herbal remedies can be traced to a time when magic and mysticism were associated with certain plants. Some folk remedies do make use of plants that contain powerful medicinal substances—some so powerful, in fact, that a misjudged dose could result in death. Consequently, most modern physicians view herbal medicine as little more than quackery.

Poisonous plants are often components of herbal remedies. Sometimes the only feature in the use of a plant that distinguishes it as medicinal or poisonous is the size of the dose. Experienced practitioners of herbal medicine can usually recognize the signs of acute poisoning, but subtle symptoms from repeated exposure to small doses of a toxic plant drug may not be so easily recognized, even by an experienced herbalist. Modern laboratory techniques are usually necessary to detect damage to internal organs such as the liver or kidneys. However, in many parts of the world, herbal medicine is the chief source of treatment for all human ailments. In one study, the World Health Organization concluded that the only way developing countries can achieve minimum health needs is to make use of traditional folk medicine.

In China, where herbal medicine has been practiced for thousands of years, there has been a fusion of folk treatments with modern methodology. Instead of treating a patient specifically for a sore knee, a liver problem, or a skin disorder, Chinese medicine prescribes for the total health of the body. A more open exchange of information between Chinese and western medicine would probably result in improvements in both.

Some Plants That Have Been Used in Folk Remedies

Buckbean (*Menyanthes trifoliata*, fig. 6.10). A perennial that blooms from April to July with white or pinkish, five-lobed, hairy flowers clustered at the top of a long leafless stem. The leaves are three lobed and grow out of the rootstock.

6.10. Buckbean (*Menyanthes trifoliata*)

6.11. Skunk-Cabbage
(*Symplocarpus foetidus*)

The stem, bearing ten to twenty flowers, usually extends above the leaves and may be one foot (30 cm) in height. Buckbean normally grows in bogs from Labrador to Alaska, south to Virginia, Missouri, and California.

Bitter plants are widely used in herbal medicine and buckbean is no exception. A bitter tea made from the leaves has been used for indigestion, for constipation, to calm the nerves, to purify the blood, and for colic in infants. Externally it has been used for cold sores and other open sores.

Purple loosestrife (*Lythrum salicaria*, fig. 7.8). A perennial that blooms from July to September with pink to red-purple flowers clustered at the top of the plant. The leaves are opposite or in whorls of three. The stem is thick, usually hairy, and may grow to 5 feet (1.5 m) high. Purple loosestrife ranges from Newfoundland to Manitoba, south to Virginia and Missouri. This plant was introduced from Europe. It is a pest that is spreading rapidly in wetlands.

A solution made from the leaves has been used, when taken internally, for dysentery and excessive vaginal discharges. Applied externally, the solution has been recommended for sore or bruised eyes and made into a salve for sores and other skin irritations.

Skunk-cabbage (*Symplocarpus foetidus*, fig. 6.11). A perennial that blooms from February to April with many small flowers enclosed in a green to purple, hood-like structure called a spathe. The spathe may grow to 6 inches (15 cm) in height. All the leaves arise from the rootstock, are prominently veined, heart shaped at the base, and may be 2 feet (60 cm) long. Skunk-cabbage grows from Nova Scotia and Quebec to Manitoba, south to Georgia.

The roots and leaves of skunk-cabbage have been used to make an herbal remedy for spasms of asthma, whooping cough, inflammation of the nasal membranes, and bronchitis. Some tribes of American Indians inhaled the odor of bruised leaves for headache. This may be an instance of a remedy that cures by distraction because after a few deep whiffs of the appropriately named plant, almost any headache would be temporarily forgotten.

Turtlehead (*Chelone glabra*, fig. 6.12). A perennial that blooms from August to September with white, sometimes tinged with pink, flowers in dense clusters at the ends of stems and branches. The flowers are two-lipped, and the upper lip arches over the lower, giving the shape of a turtle's head. The leaves are opposite with prominent midribs. The stems are smooth, often branched above the middle, and may be 4 feet (120 cm) in height. Turtlehead grows from Newfoundland to Manitoba, south to Georgia and Alabama.

6.12. Turtlehead (*Chelone glabra*)

6.13. Bugle-Weed (*Lycopus virginicus*)

In herbal medicine, the plant has been prescribed for liver problems and for intestinal worms in children. A salve made from the leaves has been used for sores, inflamed breasts, and the itching and irritation of hemorrhoids. Some tribes of American Indians used it as a laxative.

Bugle-weed (*Lycopus virginicus*, fig. 6.13). A perennial that blooms from July to September with tiny white flowers in dense clusters at the base of leaves. The leaves are opposite and narrow, with sharp teeth. Stems are square, slightly hairy, and may reach 3 feet (90 cm) in height. Bugle-weed grows from Nova Scotia and Quebec to Manitoba, south to Georgia and Texas.

In the mid 1800s, bugle-weed was reported to be mildly narcotic, astringent, and sedative. It has been recommended for coughs, bleeding of the lungs, and tuberculosis.

Edible Wetland Plants

Why Know Them?

In this high-tech era of rapid transportation and very efficient freezers, when well-stocked supermarkets are available to almost everyone, who needs to know about edible wild plants? For the purposes of survival, probably no one. Even if all transportation and freezing facilities failed and supermarkets had empty shelves, knowledge of edible wild plants would be of little value to residents of New York City, Philadelphia, Chicago, or Los Angeles. There simply are not enough edible wild plants out there to feed so many people. In the event of total failure of the supply system and electricity, millions would starve to death. When hunter-gatherers foraged for edible plants, the population was measured in tens of individuals per hundreds of square miles rather than in the millions.

The most valid reason for learning to identify edible wild plants is probably the same reason that people climb mountains: because they are there. Those who love the out-of-doors find satisfaction in being able to recognize poisonous, medicinal, and edible plants. On the practical side, it is always possible that a camper or hiker could become lost in the wilds. Knowing which plants were edible could be very helpful.

Plant Conservation

It is especially important to consider plant conservation when discussing the collection of wild food plants. The individual plants of a given species are seldom randomly distributed throughout their geographic range. Instead, they often occur in scattered clumps in those parts of the range where environmental conditions are suitable for their growth and reproduction. In collecting enough plants for a single meal, an entire local colony could be eliminated. This is less damaging when the plants are perennials and are cut at ground level leaving the rootstock to generate new plants. If the plants are annuals, collection is more damaging because young shoots, before they produce flowers and seeds, are usually the most desirable for food.

It was emphasized in chapter 1 that wetlands are among the most endangered habitats on earth. As the sizes of North American wetlands shrink, the number of individuals of each species must also shrink. Reducing this number even further by collecting some species for food in the remaining wetlands is not a practice that can be recommended. The following species are listed as edible for informational and historical interest only.

6.14. Common Arrow-Head (*Sagittaria latifolia*)

Common arrow-head or wapato (*Sagittaria latifolia*, fig. 6.14). A perennial that blooms from July to September with white flowers usually in whorls of three on a leafless stalk. The leaves are arrow-shaped and all arise from the rootstock. The leaf stalks are usually as long as the water is deep. Depending on the depths of the water, the flowering stalk may be 40 inches (1 m) long but is usually shorter. Common arrowhead grows in pond margins, swamps, and slowly flowing streams throughout the United States, southern Canada, and northern Mexico.

This plant is also known as duck potato because it produces tubers on its roots that range from the size of a pea to that of an egg. These were an important source of carbohydrates for American Indian tribes throughout North America. The Lewis and Clark expedition bought them by the bushel from local Indian tribes and used them for food while exploring the northwestern United States. They have an unpleasant taste if eaten raw but are delicious when cooked. They can be prepared in the same way as potatoes.

Broad-leaved cattail, narrow-leaved cattail (*Typha latifolia, T. angustifolia,* fig. 6.15). Perennials that bloom from May to July with tiny unisexual flowers densely crowded in cylindrical clusters at the ends of leafless stems. The male flowers are at the top of the cluster with the female flowers below them. In narrow-leaved cattail, the male and female clusters are usually separated by $^1/_2$ to $1^1/_2$ inches (1–3 cm). In broad-leaved cattail, there is usually no space between male and female flowers. The leaves form a sheath around the flowering stem at its base. In narrow-leaved cattail, the leaves are typically no more than $^1/_2$ inch (12 mm) wide. Those of the broad-leaved cattail are about 1 inch (2.5 cm) wide. The leaves and flower stalks in both species may be 8 to 10 feet (2.4–3 m) high. Cattails grow in fresh and brackish water, swamps, pond margins, and roadside ditches throughout the United States and southern Canada.

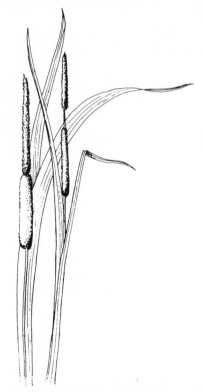

Cattails have a long history of use as food plants. The heart of the young shoots can be used in salads or as a stir-fry vegetable. The young flower spikes, before the yellow pollen is visible, can be boiled and eaten like corn on the cob. The pollen can be collected in great quantities and used directly or mixed with wheat flour to make bread. The rootstock is very rich in starch that can be extracted and used as flour. Cattails have great potential as a commercial source of food for the future.

6.15. Broad-Leaved Cattail (*Typha latifolia*) (*left*); Narrow-Leaved Cattail (*Typha angustifolia*)

Large cranberry (*Vaccinium macrocarpon,* fig. 6.16). A perennial evergreen shrub that blooms from June to August with pink flowers on long stalks that arise from the axils of smaller leaves toward the base of the stem. The flower has four petals that are bent backward with stamens forming a central, yellow, protruding cone. The leaves are alternate, with rounded tips, to $3/_4$ of an inch (17 mm) long. The stems are very long and slender, trailing on the surface of the bog with upright branches. A related species, small cranberry (*V. oxycoccus*, fig. 5.3), is very similar but with pointed leaves and flower stalks that grow from the tips of the stems. Cranberry plants grow in sphagnum bogs and on wet sandy shores from Newfoundland to Manitoba, south to Virginia and Ohio and in the mountains to North Carolina and Tennessee.

6.16. Large Cranberry (*Vaccinium macrocarpon*)

Large cranberry is the commercial cranberry that is cultivated in Nova Scotia, Cape Cod, New Jersey, New York, Wisconsin, Oregon, and Washington State. It grows wild and is harvested for local consump-

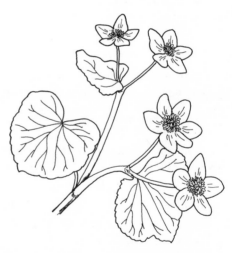

6.17. Marsh-Marigold (*Caltha palustris*)

tion in areas where there are bogs. The best time to harvest is after the first hard frost. The value of cranberry sauce to the Thanksgiving or Christmas turkey is well-known to everyone.

Marsh-marigold (*Caltha palustris*, fig. 6.17). A perennial that blooms from April to May with bright yellow flowers. The leaves are alternate, shiny green, heart-shaped at the base with rounded tips, and with long

6.18. Orange Touch-Me-Not
(*Impatiens capensis*)

stalks. The stems are hollow, often lying on the ground with upright flowering branches, to 2 feet (60 cm) long. Marsh marigold grows in marshes, wet meadows, swamps, and wet woods, from the subarctic south to South Carolina, Tennessee, and Nebraska.

The leaves of marsh-marigold or cowslip if picked before or during flowering can be eaten as a potherb if thoroughly cooked in two or three changes of boiling water. The flower buds can be cooked and pickled as a substitute for capers. Caution: the fresh plant contains a toxic substance, so it should not be eaten raw. Cooking deactivates the poison.

Orange touch-me-not or *jewelweed* (*Impatiens capensis*, fig. 6.18). An annual that blooms from June to September with orange flowers dangling from slender stalks that arise in the axils of the leaves. The flowers have reddish-brown spots and an inflated throat. The leaves are alternate, oval, coarsely toothed, on long stalks. Stems are succulent, branched, to 5 feet (1.5 m) high. Orange touch-me-not grows along the margins of swamps, in roadside ditches, and in wet woodlands from Newfoundland to Alaska, south to Florida and Texas. A related species, yellow touch-me-not (*I. pallida*), is very similar but has pale yellow flowers.

The young stems and leaves of yellow and orange touch-me-nots can be cooked as greens in two changes of water. When the seedpod is mature, it bursts explosively at the slightest touch, thus the name touch-me-not. The seeds are tasty, if tiny, snacks with the flavor of butternuts.

7

Through the Year

Causes of the Seasons

After viewing a frozen snow-covered swamp in January, then examining the flower of a pitcher plant in a peat bog on a hot day in July, it is difficult for the senses to make a connection between these contrasting conditions and the astronomical phenomena that brought them about. The fact that the earth is moving at a speed of about 18.5 miles per second around the sun seems an unreal and useless piece of information. But this is a very real and very relevant fact if we wish to understand the winter swamp and the summer bog.

The earth completes one revolution around the sun in a year of about 365.25 days. Our calendar of twelve months is organized to recognize only 365 days, so each year astronomical events gain one-fourth of a day. We solve the problem by adding an extra day to the calendar in February every four years. If we did not do so, in four hundred years, spring would occur in what this calendar indicates as January.

The earth does not go around the sun in a perfect circle, but orbits in an ellipse. Thus at its closest approach, the earth is about 3 million miles closer to the sun than at the opposite side of its orbit. But this variation in distance from the sun is not what causes summer and winter. In fact, the earth is closest to the sun during the first week in January and farthest away during the first week in July. Instead the seasons are caused by the orientation of the earth's axis relative to the plane of its orbit.

Let us imagine a great sheet of paper passing through the earth and the sun. This is the plane of the earth orbit. If the earth's geographic axis was perpendicular to this plane, forming a ninety degree angle with it, there would be no seasonal change except a little warmer temperatures when the earth is nearest the sun. The day length and the amount of solar radiation would be the same throughout the year for any spot on the earth. In this world, the rays of the sun would always be perpendicular at midday at the equator, and you would have to travel towards the pole to find a frozen snow-covered swamp or towards the equator to find a hot day. You would not find them both in the same place in the same year anywhere on earth.

Rather than being perpendicular to the plane of its orbit, the earth's axis is tilted at an angle of 23.5 degrees from that perpendicular. As the earth journeys around the sun, the northern hemisphere is tilted first toward and then away from the sun. About June 21, the north pole is inclined toward the sun, whose rays will be perpendicular to an imaginary line around the earth 23.5 degrees north of the equator called the Tropic of Cancer. Approximately six months later the earth will be on the opposite side of its orbit, and the northern hemisphere will be tilted away from the sun. On or about December 21, the rays of the sun will be perpendicular to an imaginary line 23.5 degrees south of the equator called the Tropic of Capricorn.

In between these extremes, the axis of the earth is not inclined toward or away from the sun and the sun's rays are perpendicular to the equator on March 21 and September 22. The change in angle of the sun's rays throughout the earth's revolution around the sun results in the seasons that are reflected in the astronomical calendar.

Adapting to Seasonal Changes

The tilt of the earth's poles relative to the sun may have existed since its origin from a gaseous mass or a dust cloud. Whatever the origin, there can be little doubt that seasonal fluctuations have existed throughout the evolution of modern plants. Some seasons are more favorable for growth than others. While some plants are acclimated to remarkably low winter temperatures, at other times in their life cycles the same plants are very sensitive to low temperatures. Freezing temperatures may be lethal to flowers and actively growing tissue. If plants were to survive, timing mechanisms had to evolve that would ensure growth and flowering only during favorable seasons. Day length provides a reliable clock because for any spot on earth, day length is the same on a given date year after year.

Plants make use of this information by means of a special pigment called phytochrome. When exposed to light, phytochrome changes to a biologically active form, and during darkness it changes back to the original form or disintegrates entirely. The details of how this mechanism works are not understood, but the relative amounts of the two forms of this pigment serve as reliable devices for turning on or off many physiological processes in plants. During daylight hours, there is always a greater proportion of the biologically active form of the pigment, regardless of the day length. During darkness is when the balance changes, and thus it is the length of night rather than day to which plants respond.

The plant functions that are controlled by day (night) length are initiated long before there is a need for that function. Time of flowering in many species is one of the functions that is so controlled. While temperature does not initiate these processes, it can be a modifying factor. Growth and development are manifestations of chemical reactions taking place inside the cells of plants. The rates of most chemical reactions decrease as the temperature decreases. During a very cold spring, the usual flowering time may be delayed as a result of slower chemical reactions with correspondingly slower growth rates.

Although in marsh habitats seasonal events are less distinct, in temperate swamp forests there are clear markers for the ecological seasons. The swelling of buds indicates the beginning of spring, the closure of the canopy signifies the beginning of summer, the onset of autumnal coloration marks the beginning of autumn, and the falling of leaves denotes the beginning of winter. These are the seasons to which living things respond, and they often do not coincide with the astronomical seasons. The farther one travels from the equator, the greater the disparity.

Spring

Since the seasons are cyclic, there is no beginning place for a discussion, but since spring is perceived as a time of beginning, or a time of reawakening, we can start there. Spring is usually a time for breaking dormancy. This is a biologically induced state that enables a plant species to survive the winter season or other stressful environmental conditions. Winter dormancy is associated with the presence of a growth-inhibiting hormone called abscissic acid. Before dormancy can come to an end, the abscissic acid must be eliminated, and for many species, a period of chilling is required before dormancy is broken. Although at present it is not thoroughly understood, winter dormancy is not a period during which the plant just ceases to function; it is a period of physiological activity.

It is a challenge to identify the very first sign of ecological spring in a marsh. It may be the appearance of the first shoots of cattails or other emergent plants, or the appearance of the first duckweeds on the surface of the water. In swamp and bog forests, the bursting buds on trees and shrubs are clear indications that dormancy has broken. In the northeast, ecological spring will arrive by as much as one and one-half months after March 21. The farther south one travels, the closer the beginning of ecological seasons will coincide with astronomical seasons. Ecological spring in wetlands may be later than in nearby uplands because water and water-saturated soil take longer to warm up as a result of water's high heat capacity.

An event of great importance to aquatic plants and animals occurs each spring in ponds and lakes. Water has the peculiar property of being most dense at 39°F (4°C). If it is cooled below this, it expands and becomes less dense. The coldest water rises to the surface; that is why ice forms on the tops of ponds rather than at the bottom. After the ice melts in spring, this stratification persists for a while and turbulence caused by wind blowing across the pond mixes only the upper layer. As spring progresses, the upper layer is warmed and the temperature of the water eventually becomes the same throughout. In the absence of thermal stratification, turbulence caused by the wind can bring nutrients from the bottom levels to the surface and oxygen from the surface to deeper levels. This is known as the spring overturn. Its duration depends on weather conditions, and it will continue until stratification is restored by a warm upper layer and a colder bottom layer.

Summer

According to the astronomical calendar, summer begins about June 21. On this date, the earth reaches a point in its orbit where the rays of the sun are perpendicular to the Tropic of Cancer at twelve o'clock noon. This remote astronomical event can be measured accurately to the second. The beginning of ecological seasons cannot be so precisely defined. In the swamp forest, ecological summer begins with the closure of the canopy. To visualize this, imagine yourself lying on your back looking skyward. In early spring you will see only blue sky, but as spring advances you will see less and less blue. Eventually you will see only the green undersides of leaves; the canopy has closed and summer has begun.

There does not appear to be a distinct observable biological event to mark the beginning of ecological summer in marshes. It would be convenient if the blooming of wetland species could be seen as heralding the onset of summer. The date of flowering in some species is controlled by the phytochrome system and initiated by the relative length of day and night. Others may be day neutral and bloom in response to some other stimulus. Consequently, times of flowering of wetland species are unreliable as markers for any season because they bloom from February to September. The transition from spring to summer is a gradual development of vegetation.

The summer season is a time of maximum flowering. Purple loosestrife, though considered a noxious weed by conservationists, offers a spectacular display of pink to red-purple flowers from July to September. During the same period of time, the delicate blue flowers of pickerel-weed (fig. 4.12) adorn large areas. From May to July, the flowers of native orchids such as moccasin flower (fig. 5.17), grass-pink (fig. 5.15), white fringed orchid (fig. 5.18), and rose pogonia (fig. 5.16) add beauty and color to peatlands. In flower also, but with less colorful blossoms, are the wind-pollinated species. These include the cattails (fig. 6.15) and most of the submergent species that must lift their flowers above the surface of the water.

The summer season is also a time of maximum vegetative growth. This sometimes creates problems when the growth is excessive and in areas where it interferes with or influences human activities. The species that are often the most troublesome are nonnative floating plants. They can clog waterways used for boating and fishing, interfere with hydroelectric production, block irrigation ditches, reduce production in rice farming, and reduce storage capacity in reservoirs. The following are some of the species that cause these and other problems.

Aquatic Weeds

The term "weed" is used several times in the following pages, and a note of explanation is in order. This word is a nonscientific term that usually refers to a plant growing where someone does not want it to grow. The connotation often is a plant that is troublesome and noxious. With regard to wetlands, weeds are plants that interfere with recreational boating, fishing, swimming, irrigation of fields for crops, home use for drinking and cooking, and commercial uses of water. Millions of dollars are spent each year on chemicals and mechanical harvesting for their eradication. However, the word "weed" expresses a human concept that has no meaning in the natural world.

Alligator-weed (Alternanthera philoxeroides, fig. 7.1) has clusters of flowers, each with five greenish or whitish sepals, on stalks arising from the axils of opposite leaves or from the tips of stems. It seldom produces seeds, so reproduction is entirely vegetative. The biology of seed germination for this species is unknown. It is a perennial with a creeping prostrate stem that may develop roots and produce erect branches at each node. Although it is usually a rooted emergent, it may develop floating mats of interwoven hollow stems that extend over the surface of the water and cause major problems.

7.1. Alligator-Weed
(*Alternanthera
philoxeroides*)

Alligator-weed is a native of South America, but it does not grow as aggressively or produce the problems there that it does in the southeastern United States. It grows as far north as Virginia but causes problems as an aquatic weed mainly in southern states. For example, in 1990 it occupied 151,875 acres (60,750 ha) in Louisiana. Although it causes the greatest problems as a floating plant, it also may invade shallow swamps and is able to grow aggressively and outcompete terrestrial plants in damp meadows.

Hydrilla (Hydrilla verticillata) has small flowers, white or transparent with a few red streaks, that appear from mid-June until August. It reproduces asexually by fragmentation, the production of small tubers, and bud-like structures called turions that detach and grow into new plants. It is a perennial sub-mergent species native to India, southeast Asia, and Australia, where it is usually a minor part of the aquatic plant life. In 1990 its distribution in-cluded every continent but Antarctica and South America. Its introduction into the southeastern United States has resulted in some of the most serious aquatic weed problems in the last half century.

Hydrilla is very similar to the native plant water-weed (*Elodea canaden-sis*, fig. 4.1). They are so similar that when hydrilla was first introduced into Florida waters, it was ignored because it was at first mistaken for water-weed. The chief structural difference between the two is that in water-weed the upper leaves are in whorls of three while those in hydrilla occur in whorls of three to eight. These species are also similar in that neither is a troublesome weed in its native environment. However, when hydrilla is transplanted to the native habitat of water-weed and water-weed to that of hydrilla, they grow very aggressively as weeds. That there is no known ex-planation for this indicates that there is still much to be learned about the ecology of alien plants.

Since its introduction, hydrilla has spread northward to the District of Columbia. In the recent past, it occupied 59,935 acres (23,974 ha) in Al-abama, Florida, Georgia, and North Carolina, with over 90 percent of it found in Florida. It has also been identified as a major weed species in Vir-ginia, Delaware, and Maryland because it interferes with recreational boat-ing and fishing.

Water hyacinth (*Eichhornia crassipes*, fig. 7.2) is a floating plant with bluish-purple flowers in a showy cluster at the tip of a stalk that may rise 16 inches (40 cm) above the surface of the water. The flowers are about 2 inches (5 cm) wide, funnel shaped, with six lobes, the upper one larger with a prominent yellow spot. It has oval leaves to 5 inches (12 cm) wide on stalks in-flated with spongy air-filled tissue that serve as floats. It spreads by horizontal creeping stems called stolons.

7.2. Water Hyacinth (*Eichhornia crassipes*)

7.3. Water Lettuce
(*Pistia stratiotes*)

7.4. European Water-Milfoil
(*Myriophyllum spicatum*)

Water hyacinth is a native of central and South America from Mexico to Argentina. It was introduced into southern North America about 1860. It is reported that specimen plants were give as souvenirs at an International Cotton Exposition at New Orleans in 1884. Today it is the most troublesome floating weed in the southeastern states from Florida to Texas. It is not as hardy in colder climates, but it has spread northward to Virginia and Missouri.

This plant grows as aggressively in native habitats as it does in alien ones. It has spread to most tropical and subtropical regions of the world and has become a very serious weed everywhere. Under good growing conditions, it can double its area of occupation in about thirteen days. Millions of dollars have been spent on mechanical and chemical treatments to control its spread. In the southern United States, Florida, Georgia, Louisiana, Mississippi, and South Carolina have reported a total of 219,937 acres (87,575 ha) covered with water hyacinth. As of 1990, 96 percent of this was in Louisiana. The area covered today is probably much larger. It has caused problems by clogging waterways and preventing their use for recreational use such as boating, fishing, and swimming.

Water lettuce (*Pistia stratiotes*, fig. 7.3) has flowers that are about 1/2 inch (1.3 cm) high and consist of a central column, the spadix, surrounded by a greenish-white sheath, the spathe. The spathe is constricted in the middle creating two cavities; the lower contains the pistil, the upper three to eight stamens. This structure indicates insect pollination but usually only one seed is produced. The flower is located in the center of a rosette of leaves that extend about 10 inches (25 cm) above the water and are prominently ribbed. It reproduces by stolons that give rise to many daughter plants. It has a very rapid rate of growth covering the water so completely that it blocks most of the light to lower levels and gives the appearance of a solid surface.

Water lettuce has a worldwide distribution in tropical and semi-tropical regions. It may be a native of tropical America, but at this time its origin is uncertain. In some regions, it is second only to water hyacinth in importance as a nuisance aquatic weed. Its distribution in North America is restricted to Georgia and the Gulf Coast states. Recent data show that it covered 7,440 acres (2,976 ha) in Florida, particularly in cypress swamps, and it is spreading rapidly.

European water-milfoil (*Myriophyllum spicatum*, fig. 7.4) is a submergent with inconspicuous, unisexual, wind-pollinated flowers in the axils of whorled bracts, with upper flowers staminate and lower ones pistillate. The stem is weak, rooting on the bottom, to 40 inches (1 m) long. This plant is a perennial with a creeping horizontal stem called a rhizome that often roots at lower nodes. It is not an important wildlife food plant but it is occasionally eaten by ducks, muskrats, and moose. These animals probably contribute to seed dispersal. It also spreads by fragmentation of the stem and by axillary buds that are produced year-round.

There are ten species of this genus in eastern North America. They all have very similar leaves and are sometimes difficult to distinguish even with flowers and fruits. European water-milfoil has been widely introduced into North America where it sometimes causes problems as a weed. A native species, common water-milfoil (*M. sibiricum*), is very similar to the alien species but usually has a more northerly distribution.

The states of Alabama, Florida, Georgia, Kentucky, and Tennessee have reported at total of 23,980 acres (9,592 ha) dominated by European water-milfoil. Most of this acreage is in Alabama with 12,515 (5,014 ha) and Georgia with 8,100 (3,240 ha). In some southern areas, the water-milfoil is being outcompeted and replaced by the more aggressive and potentially more harmful hydrilla.

7.5. Curly Pondweed
(*Potamogeton crispus*)

Curly pondweed (*Potamogeton crispus*, fig. 7.5) is a submergent with inconspicuous flowers in short dense clusters. It has wavy margined, alternate leaves to 3 inches (8 cm) long and $1/2$ inch (12 mm) wide. The stems are flattened, branched, to 32 inches (80 cm) long. It is a perennial with a slender underwater rhizome. Winter buds may be more important than seeds in the propagation of the species. They are bur-like, to 1 inch (2.5 cm) long, and form at the bases of leaves.

Curly pondweed was introduced from Europe into northeastern North America before 1814 and has become widespread. It may become an aggressive weed and cause serious problems in polluted water or water that has become enriched with nutrients. It is found in ponds, lakes, and sluggish streams, and sometimes in muddy, polluted, or brackish water from Quebec and Ontario to Minnesota and South Dakota, south to Oklahoma and Alabama.

Water chestnut (*Trapa natans*, fig. 7.6) is a floating annual with small, inconspicuous, white flowers of four petals. The sepals are fused into a four-lobed tube that, in the fruit, become spines or horns. The stem is rooted in soft mud and may be to 16 feet (4.8 m) long with paired plume-like structures, to 3 inches (8 cm) long, at submerged nodes. There is a crowded rosette of floating leaves at the stem tip. Each leaf is widest at the base, up to 2 inches (5 cm) tapering to a point, with an inflated leaf stalk that serves as a float. The rosette of leaves at the stem tip often breaks off and becomes a free-floating plant.

When the fruit is mature, it

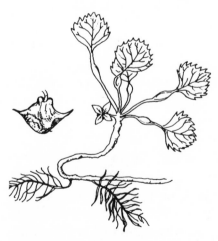

7.6. Water Chestnut (*Trapa natans*)

sinks to the bottom and spends the winter there. The seed, which is killed by air drying, is viable for one year only. Water chestnut has a worldwide distribution. It is a native of Eurasia where it is an endangered species in some regions. The nut is edible, and in India it is an important crop plant. In North America, it is a noxious weed where it has interfered with recreational boating, fishing, and swimming. It is most common in mid-northeastern North America and ranges as far south as Virginia. It has been identified as a major weed species in New York, Vermont, and Massachusetts.

7.7. Common Reed
(*Phragmites australis*)

Common reed (*Phragmites australis*, fig. 7.7) has small inconspicuous flowers in dense clusters 8 to 16 inches (20–40 cm) long at the tips of stems that may grow to 15 feet (4.5 m) in height. Flower clusters are purplish at time of blooming, from August to September, then become light brown and feathery, darkening somewhat throughout the year. It has leaves to 24 inches (60 cm) long and 2 inches (5 cm) wide alternately attached on opposite sides of the stem. In spite of the elaborate flower cluster, it produces very few seeds. It is able to spread rapidly by a fast-growing rhizome that often grows along the surface of the ground. These will sometimes grow over bare rocks for a distance of up to 25 feet (7.5 m). It has little value as a wildlife food plant but does provide cover for nesting and breeding for some bird species.

Common reed is an attractive member of the grass family with worldwide distribution. It is considered by some botanists to have a wider geographic distribution than any other plant species. Unlike many species that become weedy, it is not an alien but a native species that does well in areas that have been disturbed by human activities. These are so numerous that it is expanding rapidly in some habitats to the exclusion of all other species. It is commonly observed in roadside ditches in the northeast probably because, unlike other freshwater wetland species, it is favored by a tolerance for the salt that washes from road surfaces. Some newspapers in that region have referred to it as ditchweed.

7.8. Purple
Loosestrife
(*Lythrum salicaria*)

The extensive lowlands along the Hackensack River in New Jersey, the so-called Meadowlands, that were at one time occupied by a variety of marsh species, have undergone modifications in water level and are currently dominated by common reed. In some parts of coastal New England, tide gates were constructed to control the daily tidal flooding. In those areas, the salt-marsh grasses were replaced by common reed. In pre-gate days, it could not survive there because it is intolerant of the daily fluctuations in water level caused by the tides.

This plant is widely distributed in North America. It has been identified as an aquatic weed species for most of northeastern North America. In recent years, it has spread to the southern states and covers at least 5,448 acres (2,179 ha) in Florida and 3,543 acres (1,417 ha) in South Carolina.

Purple loosestrife (*Lythrum salicaria*, fig. 7.8) has pink to red-purple flowers in a crowded, elongated cluster at the tips of stems and branches. The leaves are opposite or in whorls of threes on a stem that may be 5 feet (1.5 m) in height. It is a perennial with three types of flowers relative to lengths of stamens and styles. Some flowers have long stamens with metallic green pollen and short stamens with yellow pollen. The species is self-incompatible and cross-pollination is effected by bumblebees, honeybees, and butterflies. This is an aggressive nonnative species that invades marshes, wet meadows, pond margins, and flood plains often to the exclusion of all other species.

Purple loosestrife was already established in North America by the 1830s. No one knows for sure how or when it arrived on this side of the Atlantic Ocean, but there are several possibilities. Seeds may have arrived in the ballast of sailing ships. Rocks and sand were commonly used for this and they were disposed of in American ports. The seeds may have arrived in hay fed to imported livestock or in shipments of wool. It is possible that the colonists brought the seeds to plant as ornamentals or medicinal herbs. The plant probably had multiple introductions, so it could have arrived by all of these ways and others.

In North America, purple loosestrife commonly grows in pure stands that may cover large areas. It has been observed that these may remain vigorous for at least twenty years. In Europe, its native land, it grows as an occasional associate in mixed-species stands. Only in extremely disturbed sites does it establish pure stands, and these lose their vigor after a few years. The reason this plant is able to sustain vigorous growth in pure stands in North America is unknown. It has been suggested that more adaptive and vigorous forms may have appeared in those populations. The behavior of this species highlights the need for more information on the ecology of alien species.

The range of purple loosestrife today includes Newfoundland and Quebec to North Dakota, and south to Virginia, Missouri, and Kansas. In recent years, its rate of spread seems to be increasing. From 1900 to 1940, its rate of spread was 297 square miles (722 sq. km) per year, while from 1940 to 1980 the rate of spread was 445 square miles (1,158 sq. km) per year. One of the reasons for this is that in spite of being listed as a noxious weed in twenty-four states, it is still being sold as an ornamental plant in nurseries across the country.

Methods of controlling the spread of purple loosestrife such as pulling, mowing, plowing under, burning, and the use of herbicides have been ineffective or excessively expensive and time-consuming. Several years ago, the Cooperative Fish and Wildlife Research Unit at Cornell University with the cooperation of the New York State Department of Environmental Conservation initiated a program of biological control. Three insects—a root-boring weevil and two leaf-eating beetles, adapted specifically to feed on purple loosestrife—were introduced from its native habitat in Europe. The ultimate goal was the elimination of 75 to 80 percent of the purple loosestrife in North America. From 1994 through 1999, the insects were released in sample plots in western New York wildlife management areas. It is too early to determine the full impact of this project, but to date the results are favorable.

Common, or broad-leaved, cattail (*Typha latifolia*, fig. 6.15) has tiny inconspicuous, unisexual flowers, densely crowded in a cylindrical terminal cluster. Male flowers are at the top of the cluster with female flowers beneath, usually with no space between them. The mature flower cluster, the "cattail," is about 1 inch (2.5 cm) in diameter. Leaves are bluish-green, nearly flat, and up to 1 inch (2.5 cm) wide. The flowering stem and leaves may grow to 10 feet (3 m) in height. Common cattail is similar in general appearance to narrow-leaved cattail. In the latter, the flower cluster is about $1/_2$ inch (12 mm) in diameter, with usually a separation of about 1 inch (2.5 cm) between male and female flower culsters, and the leaf is typically no more than $1/_2$ inch (12 mm) wide.

Common cattail is a native of North America with a thick creeping rhizome. It blooms from May to July and each flower cluster may produce 117,000 to 268,000 seeds that have parachutes of hairs for dispersal by wind. In order to germinate, the seeds require a moist or wet substrate, warm temperatures, and long day-short night exposure. Under experimental conditions, a single seed planted April 1 grew 98 aerial stems and 104 lateral shoots by November 1. Seeds provide a means for invading new areas, but it is an extensive rhizome system that maintains and expands existing stands.

This species will grow vigorously anywhere the soil stays wet, saturated, or underwater for most of the year. It does not survive if the water is more than about $2^1/_2$ feet (80 cm) deep. Along a gradient, it grows best in water that is shallower than for bulrush (*Scirpus sp.*) and deeper than for common reed. Although it is mainly a freshwater wetland plant, it sometimes grows in brackish marshes.

The range of common cattail in North America is from central Alaska and northeast Canada to Newfoundland and south through every province and state to Mexico and Guatemala. It invades new habitats readily and grows aggressively in wet meadows, marshes, pond and lake margins, irrigation canals, and streambeds and backwaters of rivers and creeks. It is one of the most troublesome aquatic weeds in the United States and Canada. It has been introduced in Australia where it does not exhibit the aggressive growth that it does in its native habitats. It apparently does not set viable seeds in that country.

Autumn

Autumn begins on September 22 according to the astronomical calendar. On this date, the sun's rays are once again perpendicular to the equator, and the length of daylight and darkness are the same all over the world. This date is closer to the midpoint of ecological autumn that to its beginning. By the middle of August, the shortening length of nights has already changed the ratio of phytochrome pigments significantly.

In woody plants of swamps and swamp forests, physiological processes are initiated that begin to close down their summer activities. Substances are produced that promote the onset of dormancy and the beginning of a special layer of cells called the abscission layer between the leaves and their points of attachment to the branches. Autumn, then, is a very active time, chemically, for wetland plants. As you travel through a wetland in autumn, you will not be aware of this chemical activity, but you will be able to see its earliest signs, the beginning of autumnal coloration. This signals the beginning of ecological autumn, and in northeastern North America it may begin as early as late August rather than in late September.

As autumn progresses, the special layer of cells developing at the base of the leaf first interferes with, then finally cuts off entirely, the supply of water to the leaf. As this is happening, the green chlorophyll pigments disintegrate revealing the orange and yellow pigments that were previously hidden by the chlorophyll. In the warm sunny days and cold nights of autumn, red pigments are manufactured that augment the oranges and yellows. This is the most colorful time of the year for woody plants of wetlands.

Flowers of Autumn

Autumn is a time of flowering for many wetland species. In some of these, blooming is controlled by length of night and the phytochrome system. Other species may be day neutral with flowering controlled by some other factor. The following are a sampling of species that are commonly seen in bloom in late summer and autumn.

Bristly aster (*Aster puniceus*, fig. 7.9) has numerous flower heads, each about 1 inch (2.5 cm) wide with vi-

7.9. Bristly Aster (*Aster puniceus*)

olet-blue rays. The leaves are alternately attached, narrow, tapering at the base and clasping the stem. It may grow to 8 feet (2.4 m) high, but it is usually smaller. Bristly aster is a highly variable native species that hybridizes freely with one or more other species. Forms are occasionally seen with lilac, pink, or white ray flowers. It is found in swamps, swales, damp thickets, and other moist places from Newfoundland to Saskatchewan, south to Georgia and Alabama.

Turtlehead (*Chelone glabra,* fig. 6.12) is a perennial with a fibrous root system. Pollination is usually accomplished by bumblebees, which are large enough to force open the lips of the flower. After a visit to one flower with a rich supply of nectar, they search widely for flowers of the same type. The roundish seedpod splits from the top to release many black seeds, each bordered by a gray corky or spongy wing that aids in dispersal by wind or water.

Orange touch-me-not (*Impatiens capensis,,* fig. 6.18) eliminates the possibility of self-pollination by having stamens that mature and are shed before the stigma of that flower becomes receptive to pollen. It is an important source of food for the ruby-throated hummingbird who contributes to its cross-pollination. The common name comes from the color of the flower and a seed capsule that at maturity bursts open at the slightest touch. The capsule splits from the base into five sections that forcefully roll inward, hurling the seeds for a distance of up to 8 feet (2.4 m). This plant is edible, and the water from cooking it is said to prevent poison ivy rash if applied immediately after exposure.

Great lobelia (*Lobelia siphilitica,* fig. 6.8) is a perennial with short basal sprouts. The specific name refers to the reputed use of the

7.10. Climbing Hempweed (*Mikania scandens*)

root by the Mohawk Indians as a treatment for syphilis. It is found in swamps, along stream and pond margins, and in open wet woods from Maine to Manitoba, south to North Carolina and Texas.

Climbing hempweed (Mikania scandens, fig. 7.10) has white to pinkish flower heads in stalked clusters from the axils of leaves. The leaves are triangular, somewhat like those of morning glory, attached in pairs on a twining stem often forming mats on other vegetation. This native plant has spread to tropical regions throughout the world. It is an unusual plant because it is the only climbing member of the aster family in North America. It is found along the margins of swamps, on stream banks, and in thickets bordering salt marshes along the Atlantic coastal plain from Maine to Florida and Texas and inland to Ontario, Michigan, and Missouri.

Swamp Rose (Rosa palustris, fig. 3.5) has crimson pink flowers to 3 inches (7.5 cm) wide, with five petals, solitary or in small terminal clusters. It has alternately attached, pinnately compound leaves, usually with seven leaflets. This is a shrubby perennial that may grow to over 6 feet (2 m) in height. As autumn advances, the flowers are replaced by bright red globular fruits or "hips" that contain numerous seeds. These remain on the plant throughout the winter months and are an important source of food for several species of birds and mammals. Swamp rose is found in swamps, marshes, on stream banks, and in other low places from Nova Scotia to Minnesota and south to the Gulf of Mexico.

Cutleaf-coneflower (Rudbeckia laciniata, fig. 7.11) has six to sixteen lemon yellow, drooping petals (ray flowers) with a dome-shaped greenish-yellow central disk. It usually has several flower heads at the ends of stems and branches each up to 4 inches (10 cm) wide. The basal leaves are long-stalked, pinnately divided, with five to seven irregularly lobed or toothed leaflets. The upper leaves are smaller, three to five parted on a stem that may

7.11. Cutleaf-Coneflower
(*Rudbeckia laciniata*)

reach 10 feet (3 m) in height. It is found in marshes, swamps, wet meadows, and roadside ditches from Quebec to Manitoba and Montana, south to Florida, Texas, and Arizona. A cultivated double form is called golden glow.

Fruits of Autumn

Autumn is also a time of fruit and seed dispersal. The colorful flowers of spring and summer attract insect pollinators. Similarly, the colorful fruits of some wetland plants attract animal consumers who serve as agents of seed dispersal. Some of the seeds that are ingested with the fruits pass undamaged through the digestive tract and are deposited with the feces of the animal. The nutritious pulp of the fruits provide the same type of reward for fruit eaters as the nectar of flowers for insect pollinators. The bright autumnal coloration of foliage in swamp forests does not have a known adaptive value. It appears to be simply a by-product of the metabolic changes leading to winter dormancy. This does not appear to be so of the colorful autumn fruits. Bright colors make the fruits stand out and thus easier to find for hungry fruit eaters.

Species without colorful fruits have developed other strategies for seed dispersal. One adaptation is for dispersal by air currents. The tiny one-seeded fruits of cattails and bristly asters have parachutes of hairs that carry the seeds to new areas. The bur-marigold (*Bidens cernua*) has fruits with four barbed awns that hitch a ride by attaching to fur, feathers, or the clothing of field naturalists. Plants such as turtlehead and fragrant water-lily (*Nymphaea odorata*) have structures on the seeds that facilitate dispersal by water.

In the above examples, evolution has produced specific structures that have no function other than seed dispersal. In some species there are no obvious adaptations for this activity. For example, purple loosestrife's numerous small seeds simply fall to the ground, yet it is spreading at an increasingly rapid rate. Actually, this small size itself may be an adaptation for dispersal because the seeds are easily carried in the mud on the feet of aquatic birds as they range from one wetland to another.

The Autumn Overturn

During the summer months, the water of lakes and ponds becomes stratified with regard to temperature. Colder water, being more dense, sinks to the bottom, and warmer water remains in the surface layers. Turbulence caused by wind passing over the surface influences water only in the upper warmer layers with no mixing of the upper and lower layers. In northern areas, as the water cools in late autumn, it eventually reaches a temperature of 39°F (4°C) throughout. This is the temperature at which water is most dense. Wind-induced turbulence will now mix all levels freely, bringing bottom nutrients to the surface and carrying oxygen to deeper levels. This important event in the life of a pond is called autumn overturn. Depending on local weather conditions, the overturning may last for several weeks. As the temperature of the water drops below 39°F (4°C), it becomes less dense, so the coldest water is always on the surface where it eventually freezes to seal the pond for the winter.

Winter

Getting Ready

When the earth reaches the point in its orbit that is closest to the sun, the northern hemisphere is tilted away from it. This occurs about December 21, and at this time the sun's rays are perpendicular, at noon, to the Tropic of Capricorn. In the northern hemisphere, this is the shortest day of the year, and the calendar identifies it as the beginning of winter. It is definitely not the beginning of ecological winter. To understand this we must revisit the swamp forest at the peak of autumnal coloration.

As the water supply to the leaf is cut off, the cells die and the bright autumnal coloration fades. In the wind and rains of October and November, the leaves break away along the abscission layer that started forming in mid—to late August. Soon the trees will be bare and ecological winter will have begun. The farther north one travels, the earlier ecological winter begins. In many parts of the northeast, the leaves are on the ground and winter has begun by October 31. In these places, December 21 (winter solstice) does not mark the beginning of any biological event. By this time, seeds have been dispersed and the trees are in full dormancy. For those who do not live in the northeast, ecological winter may begin later than October 31.

Wetland plants begin to get ready for winter long before the leaves have dropped. A change in the ratio of phytochrome pigments results from a progressive lengthening of the period of darkness in late summer. Compounds manufactured in the leaves of woody species enhance the readiness of plants for winter. Before conductive tissue is blocked by the abscission layer, useful substances such as phosphorus, nitrogen, and carbohydrates are transported from the leaves to permanent tissues of the stem or to underground storage organs. The hormone abscissic acid is formed in the leaves and transported to other parts of the plant. One of its functions is as an inhibitor of growth. Cessation of growth is essential in order to survive the winter. These and other physiological processes promote the onset of dormancy. Similar changes take place in herbaceous plants; annuals become ready to spend the winter as dormant seeds, and perennials as underground rootstocks.

Winter Aspect

The winter aspect of many wetlands is a study in shades of brown. On display in marshes are the flower stalks of cattails with their brown cylinders of seeds, often with fluffed sections caused by wind and seed-eating birds. In other areas, the tan-to-brown flags of common reed control the scenery. Both of these species may persist throughout the winter and into the next growing season. In swamps, the brown seed balls of buttonbush (*Cephalanthus occidentalis*) on their long drooping stalks sway in the breeze. The bright red hips of swamp rose, the red fruits of spicebush (*Lindera benzoin*), and the red berries of winterberry (*Ilex verticillata*) add touches of color to the scene.

Black spruce and evergreen ericaceous shrubs give winter color to northern bogs. Evergreenness, however, is more that a pleasant interlude for the eye; it serves an important function in these plants. Unlike deciduous plants, the leaves are not shed each year. Evergreen leaves continue to function for up to five years. The deciduous lifestyle requires the expenditure of large amounts of nutrients in the production of a new crop of leaves each year. In getting more than one year of photosynthetic activity from each leaf, the evergreen lifestyle is more efficient. This is of particular importance in habitats such as bogs that are poor in mineral nutrients.

8

Naming, Collecting, and Preserving Plants

Plant Names

There are more than 300,000 known species of plants on earth today. Each of these has a name, and some have more than one. When a new species is discovered, the individual who recognized it as new has the honor of giving it a botanical name. Every species thus has a Latinized botanical name. It is in Latin because this language is no longer spoken in any culture, and so it will never change. As a result, the Latin name will have the same meaning five hundred years from now as it does today. In addition to the botanical name, many plants have one or more common names that usually date to antiquity. These were given by people who were familiar with the plant because it was harmful to humans in some way, was a source of medicine, was useful for food, or had outstanding physical characteristics.

Marsh marigold is a common plant of marshes, swamps, and wet meadows. It is also found in Europe where it has a name in the language of each country in which it grows. It may even have more than one name in each language. For example, in the United States marsh marigold is also known as cowslip, king-cup, may-blob, palsy-wort, and verrucaria. In addition, sometimes the same common name is applied to entirely different plants. The primrose genus (*Primula*) and bluebells (*Mertensia virginica*) are also sometimes referred to as cowslips.

In order for a plant name to be of scientific value, it must be the one and only name that universally refers to that plant. If a botanist wishes to publish a paper on research conducted on marsh marigold, the plant name must be recognized by other botanists. One of the American common names would mean nothing to botanists in Russia, India, or Germany. This is why research reports always identify plants by their botanical names. The botanical name for marsh marigold is *Caltha palustris*, and it is spelled the same way in every language of every country in the world. No other plant on the globe has this name.

Common or Folk Names

While every known plant species has a Latinized botanical name, most do not have common or folk names. Possible reasons for this are that many plants are small and escape notice, or are growing in areas infrequently visited by humans. For example, mosses and liverworts are inconspicuous low-growing plants collectively referred to as bryophytes. There are more than 22,000 different species of these plants, some of which are difficult to distinguish even by professional botanists. Among the flowering plants, some closely related species are so similar that an entire cluster of species, called a genus, may be known by a single common name.

Although common names are unsuitable for the identification of plants in research papers, they represent a wealth of information and folklore. Most common names are descriptions of some feature of the plant. For example, arrowhead is a name that refers to the shape of the leaves. Even if there are two plants that have this name, no one will be surprised to find they both have leaves shaped like arrowheads. Common names frequently refer to color; it is easy to visualize the flower colors of purple loosestrife or rose pogonia. Milkweed is a wetland plant with an abundance of milky white sap.

Other descriptive characteristics featured in folk names are odor, habitat, and geography. Skunk-cabbage is an early blooming wetland plant with large leaves that, when broken or bruised, have a skunk-like odor. The bright yellow flowers of marsh marigold and swamp candles will be observed only in wetlands. Virginia chain-fern was so named not because Virginia is the only place it occurs but because this is the place it was first observed or collected.

One of the most common themes for folk names is the medicinal uses of plants. Names such as itchweed and palsywort leave little doubt about the medicinal condition to which they refer. A sixteenth-century physician named Paracelsus popularized a concept known as the doctrine of signatures that had a great influence on the naming of plants. According to this concept, the creator gave each plant species a characteristic to indicate its use for humans. None of the treatments suggested by the doctrine of signatures have been substantiated by modern medicine.

Other common names are intriguing, colorful, and sometimes ominous. Beaver poison is another name for the deadly water hemlock, a single bite of which can kill a human in fifteen minutes. Dragon's mouth, death camass, mad-dog weed, and snake head are thought-provoking names that have been applied to other wetland species. Devil's guts and hellweed are names farmers call dodder, probably to express their displeasure with a plant that is sometimes parasitic on crop plants.

Botanical Names

Botanists have been using Latin to name plants for hundreds of years. Before the mid-eighteenth century, these names often consisted of several words and were more like descriptions than names. In 1753 a Swedish botanist named Carl Linnaeus wrote a book entitled *Species Plantarum*, which, freely translated, means "The Species of Plants." In the book he used a two-word system to give names to all the plants in the world known to him. Although it met with resistance from some botanists of the time, this method greatly simplified the naming of plants. This two-word or binomial system is used by botanists today.

There is a well-defined procedure for giving a newly discovered plant species a name. Following the system initiated by Linnaeus, the botanical name consists of a generic name and a specific name, designating membership in a genus and species. The botanists coming after Linnaeus added a third component to the name: the initials of the botanist who named the plant. All the plants named by Linnaeus have an L. following the specific epithet. For example, the botanical name for marsh marigold is *Caltha palustris* L. This plant is in the genus *Caltha*, its specific name is *palustris*, and it was named by Linnaeus. In nontechnical publications, the initials of the botanist who named the plant are often omitted. In writing the botanical name, the genus is always capitalized and the specific epithet lowercased. The word *species* is both singular and plural as illustrated in the following sentence. "The genus *Caltha* has at least two species but some genera have only one species."

The botanical name provides both descriptive information about the species and information on its evolutionary relationships. All the species that make up a genus evolved from a common ancestor, thus, a genus is a group of closely related species. Likewise, a group of closely related genera make up a family and a group of similar families is an order. All of the genera in the lily family evolved from a common ancestral genus and are more similar to one another than they are to the genera of any other family. Most generic names are hundreds of years old and are derived from ancient Latin or Latinized Greek words.

The second word of the botanical name is the specific epithet. It is usually a word that describes some characteristic of the species. For example, *Quercus bicolor* is the botanical name for swamp white oak. *Quercus* is the name used in ancient Rome for oak and *bicolor* refers to leaves that are green on top with white hairs on the underside. *Typha* is the genus name for cattails and *latifolia* means broad leaved. *Typha latifolia* then, is broad-leaved cattail. It should be noted that the botanical name for this plant must include both the genus and the specific epithet. Latifolia is the specific epithet for several plants. Only when it is used with *Typha* does it mean broad-leaved cattail. This is true for all specific epithets: they are parts of the botanical names only when used with genus names.

Although botanical names are indispensable for professional botanists, for the uninitiated they sometimes seem long and difficult to pronounce. *Ranunculus trichophyllus* is the formidable botanical name for the water-crowfoot. Many—perhaps most—botanical names are shorter than this, and with experience and a little effort they become much easier to use. One of the pleasures of being familiar with plants is having the ability to talk with others who have similar interests. For maximum communication with others, at all levels, learning the botanical as well as the common name is recommended.

What Is a Species?

Species is a term that has been used frequently in the preceding pages, so a few words of explanation are appropriate. The species is the basic unit of classification. What this means is that the living representative of a genus, a family, or an order is a species. The genus, family, and order are concepts but the species can be seen and touched.

A species is a group of plants that resemble one another more than they do members of other species. The plants in a species interbreed freely but do not interbreed with members of other species. Although these statements are generally accepted as reliable descriptions of a species, they are oversimplifications because sometimes different species do interbreed to produce hybrids. These hybrids are usually but not always sterile and do not produce offspring. To complicate matters even more, some species develop seeds without pollination and the subsequent union of male and female sex cells. These seeds germinate and grow into plants that are the exact replicas, or clones, of the parent plant. Every student of botany soon learns that it is difficult to formulate a definition of a species that does not have exceptions. The reader is challenged to explore this topic further in the readings at the end of this book.

Collecting Plants

Humans are collectors of the things that interest them, from bottle caps to vintage cars. It is not unusual, then, that people who are interested in plants should collect plants. It is likely they are individuals who love the out-of-doors. Collecting plants not only satisfies their collecting desires but also provides exercise and fresh air. It is a pleasurable activity, but one that should be pursued with some caution.

Where to Collect

Plant collectors do not have the freedom to collect all the plants they want wherever they see them. Most of the land surface in the United States is owned by someone or some organization. To avoid legal entanglements, permission should be acquired before collecting on private property. Collecting is prohibited on some municipal, county, state, and federal parks, but limited permission can sometimes be granted if park managers are approached with tact.

Where and What Not to Collect

Even after permission has been granted to collect in a particular area, the collector is not absolved of all responsibility. Out of consideration for environmental conservation, it is a good practice to follow a few simple rules of conduct. When there are only a few plants of a species growing in an area, it is best to collect where they are more abundant. If there is only one plant of that species growing there, it should *never* be collected. The collecting area should be altered as little as possible by the collector. Collecting the last specimen of a species eliminates the colony from that area.

State conservation departments can provide lists of rare and endangered plants for their states. To avoid extinction of these species, every effort should be made to assure their survival and perpetuation in the natural world. They should not be collected. A suggested alternative to collecting these plants is to collect their seeds and grow your own. Mature seeds can be harvested without damaging the plant, and it will be a challenge to try to create environmental conditions under which they will germinate and grow.

Plant collectors should be aware that some plants are poisonous and may cause serious skin rashes on individuals sensitive to them. The most common hazards are poison ivy and poison sumac. Poison sumac is a shrub with leaves somewhat like staghorn sumac. It may grow to a height of 15 feet and is found mostly in swampy areas. Poison ivy, poison oak, and western poison oak are closely related species. They all have similar leaves and cause similar reactions in sensitive individuals. About 80 to 85 percent of the population in the United States is allergic to these plants. Even if you are not allergic, you should avoid them because repeated exposure sometimes causes sensitivity to develop.

In both poison sumac and poison ivy, a skin rash develops directly after contact with the juices of these plants. Some plants cause what are called phototoxic reactions. When sensitive individuals have the juices of these plants on their skins and are then exposed to sunlight, a sunburn-like rash develops within twenty-four hours with blisters in forty-eight hours. Sometimes even after the rash has disappeared, skin discoloration remains. Fortunately most people are not sensitive to these plants, and most plants do not contain the compounds that cause phototoxic reactions.

Tools for Collecting

Experienced collectors usually have a kit that contains the essential tools. It may be stored in a backpack or in the trunk of a car. The kit should be readily accessible at all times because good plant specimens are sometimes spotted when not on specific collecting trips. The basic items needed in the kit are a cutting tool, containers for the specimens, notebook and pencil or pen, identification tags, and a hand lens. These are described in more detail below.

Cutting Tool

Every collector needs a cutting tool such as a penknife or a pair of hand clippers or pruning shears. Any kind of pocket knife with a sharp blade will be satisfactory, but sometimes for woody plants, hand clippers are better.

Plant Containers

The traditional type of container used by professional botanists for plant collecting is called a vasculum. It is usually constructed of a light metal such as aluminum, with an easily opened and closed lid, and equipped with a shoulder strap. When the vasculum is lined with wet newspaper, it will keep plants from wilting for several days. These containers can be purchased from biological supply houses but are rather expensive. A selection of scientific supply houses are listed at the end of this chapter.

Plastic bags are less expensive, easier to store and transport, and even many professional botanists are finding them more convenient than vascula. For smaller plants, bags that have a zipper-style closure work very well. Larger bags that are closed with a twist-tie can be used for larger specimens. Bags of several different sizes, from sandwich size for small plants to very large ones, should be included in the collecting kit. Experience will teach the best mix of sizes to have available.

To keep collected samples from wilting, a piece of wet newspaper can be placed inside and the bag should be kept out of direct sunlight. Specimens prepared in this manner will remain fresh for two days or more. If the plastic bags or vascula are stored in a refrigerator, the specimens will remain in good condition for up to a week. Under no circumstances should plants be frozen if they are to be pressed or dried. When frozen specimens are thawed, they appear to have been cooked.

Notebook

The importance of a field notebook cannot be overemphasized. A record of each species collected should be made on the spot if possible. A collecting trip may yield several species. If recording the data on these is postponed until the end of the day, details may be forgotten or the collection site of one species may be confused with that of another. The data recorded for each plant collected should include the habitat: dry, sunny hillside; moist, shady woods; margin of a cultivated field; edge of a swamp; and so on. The geographical location should be noted with as much detail as possible. If U.S. Geological Survey topographical maps are available, rural roads, wetlands, fields, and forests will be identified and the latitude and longitude can be determined. For information on USGS topographical maps and how to acquire them, write to:

United States Geological Survey
Map Distribution
1200 Eads Street
Arlington, VA 22202

Knowing the exact location of the site and the date of collection are important if the collector wishes to return at another season for flowers or fruits. This information can also be helpful to other collectors.

Other items that should be recorded at the time of collection, since they may change with time, are flower color and odor. The number of flower petals should be noted because some may fall after the plant is placed in a collecting bag.

Identification Tags

Each specimen collected should be identified with a number or letter. Suitable tags can be inexpensively purchased at almost any store that sells office supplies. The field notebook entry should be listed under this number and a tag should accompany the specimen at all times. It can be attached directly to the plant or placed in a bag with only one specimen in it. The identification number or letter on the tag should be written with a pencil or a pen that does not smudge or smear when moistened.

Hand Lens

An item that may not be essential but can be very useful to the collector is a small hand lens. One that magnifies about ten times is sufficient for most uses. The hand lens is especially useful when examination of flower parts is necessary for the identification of a species.

The Specimen

Herbaceous Plants

When collecting herbaceous plants, it should be kept in mind that the single most important features are the flowers because they are necessary for identification. If the plant is to become part of a collection, it should be in bloom when it is collected. Sometimes identification is easier if both flowers and fruits are available. Some plants bloom over a period of time so that both flowers and fruits can be collected on the same specimen. Usually, though, if fruits are required, it will be necessary to return to the collection site later in the season. Since the flowers are essential for identification, if the plant is unknown to the collector it is useful to collect a few extra. This will allow for the dissection that is often necessary for identification without damaging the specimen that will be saved for the collection.

The ideal specimen should be one that is representative of the species. It should not be the largest or smallest plant in the colony but rather near the size of most plants of that species at that location. The specimen should be in good physical condition with a minimum of insect damage. There should be enough leaves to clearly demonstrate whether they are attached to the stem in pairs (opposite) or singly (alternate).

Sometimes for smaller plants the entire specimen, including the roots, can be taken. These should be removed carefully so as not to damage or deface the collecting site. When a collector leaves a collection area, it should look exactly the same as before the plants were taken. For some larger plants, usually only the upper portion of the stem with its leaves and flowers is collected. In addition to a leaf-bearing stem, some plants also have leaves, called basal leaves, that grow directly from the rootstock. Whether they are the same or different from the stem leaves is sometimes an important identifying feature. When plants have basal leaves, a few of these should also be collected.

Ordinarily one specimen of a species is enough for most collectors. If the species is less than abundant, the collector should be guided by good conservation practices and limit the number of samples to one. When a species is plentiful, two complete specimens may be taken in case one becomes damaged. One specimen can sometimes be used for confirmation of identity by sending it to an expert botanist. It is usually not necessary or a good plant conservation practice to take more than two specimens. These statements are especially appropriate for wetland species since wetlands and the total number of individuals of wetland species are shrinking.

Woody Plants

Many trees and shrubs can be identified by leaves alone, but some require fruits or seeds. For example, the color of mature fruit is helpful in distinguishing among species of dogwood. Acorns are essential in the identification of species of oaks. The specimen for a tree or a shrub should consist of a twig from the end of a branch with enough leaves to clearly show leaf arrangement. Identification almost always requires knowledge of whether the leaves are opposite or alternate. The end of a branch that developed in sunlight is best for this purpose. A twig growing in the shade grows so slowly that the distance between alternate leaves may be so short as to appear to be in pairs or whorls.

The central part of the stem is called the pith. The appearance of the pith in a twig is sometimes used in identification. To expose this part, use a razor blade or a sharp knife to split the cut end of the twig and excise the upper half of the split section.

Identifying the Plant

If a plant that has been collected is unknown to the collector, it is easier to identify as a fresh specimen than as a pressed and dried one. It is advisable, then, to identify the plant as soon as possible. Most plant manuals and handbooks include dichotomous keys for the identification of unknown species. Dichotomous keys are based on the assumption that any collection of plants can be divided into two groups by an observable characteristic that is present in one group but not in the other.

When comparing two specimens of the same species, there can be much variation in physical characteristics. For example, two plants grown under different environmental conditions may be different in height and thickness of stems and the number, shape, and size of leaves. In these same two plants, though, there will be very little variation in flower structure. In most instances these are observable with the naked eye, but sometimes a simple hand lens is helpful.

The use of a dichotomous key can best be illustrated by a small group or plants. Consider a collection with the characteristics listed below.

Plant 1: 10 white petals, 5 stamens, more than 1 pistil
Plant 2: 5 white petals, 10 stamens, 1 pistil
Plant 3: 3 white petals, 6 stamens, 1 pistil
Plant 4: 6 blue petals, 6 stamens, more than 1 pistil
Plant 5: 5 blue petals, 5 stamens, more than 1 pistil
Plant 6: 3 blue petals, 3 stamens, 1 pistil

A dichotomous key for these plants is given below.

A. Plants with petals and stamens in numbers divisible by 4 or 5
    B. Petals blue . . . . Plant 5
    B. Petals white
        C. Pistil 1 . . . . Plant 2
        C. Pistils more than 1 . . . . Plant 1
A. Plants with petals and stamens in numbers divisible by 3
    D. Petals white . . . . Plant 3
    D. Petals blue
        E. Pistil 1 . . . . Plant 6
        E. Pistils more than 1 . . . . Plant 4

In this key, the contrasting statements are given in uppercase letters. The user must repeatedly choose one characteristic over another in progressing to the identity of the unknown plant. Two observations are in order for this type of key.

(1) Each of the contrasting statements from which the user must choose are the same number of spaces from the left margin.

(2) The contrasting statements often begin with the same word followed by a word or statement that expresses a different condition. For example:

B. Petals blue
B. Petals white

There are other ways that dichotomous keys can be organized, but they all require a series of choices between characteristics to arrive at the identity of an unknown plant. The above key is an oversimplification because it involves a very limited group. The keys in plant manuals are much more complex because they cover a greater number of plants.

Books on plant identification are listed in the bibliography.

Preserving the Collection

A herbarium is a collection of pressed, dried, and mounted plant specimens. The objective of collecting plants for most amateur botanists is to accumulate a personal herbarium. This differs from the professional botanist only to the extent that the latter usually collects for an institution such as a college, university, or botanical garden.

The population of the United States is increasing at the rate of nearly two million people per year. With the passage of time, wetland areas already under stress from human activities will become even further disturbed, if not entirely eliminated. Many plants that are abundant today will without doubt become much less so in the future. The amateur's collection may thus become a valuable documentation of rare, endangered, or extinct plants.

In order to be useful, a plant specimen must be properly prepared and include essential collecting information. A method that has been successfully used by botanists for many years is to press the specimen flat, let it dry, then attach it with glue to a sheet of white paper. This procedure is the same for both amateurs and professionals. Mounted in this manner and protected from insects, the specimen will last for hundreds of years.

Equipment Needed for Pressing

Collectors will develop individual routines for preparing specimens, but some necessary equipment includes a plant press, corrugated cardboard ventilators, blotters or newsprint, mounting paper, labels, and an adhesive. These items and their uses, with some alternatives, are explained below.

The Plant Press

The function of the plant press is to thoroughly flatten the specimen and hold it that way until it dries. This is accomplished by placing the plant between two solid or wooden grid frames that distribute the pressure evenly. The frames are held together with ropes or straps that can then be tightened to the desired amount. Plant presses can be purchased at biological supply houses or be constructed inexpensively. Two pieces of 1/4-inch plywood or perforated masonite, each 12 by 18 inches, will serve satisfactorily as frames. Two pieces of heavy cord or, preferably, canvas straps with buckles, each about five feet in length, can be used to hold the frames together.

Corrugated Cardboard Ventilators

Cardboard ventilators can be purchased from biological supply houses or they can be made from corrugated cardboard boxes. Ventilators cut from boxes should be 12-by-18 inches with the corrugations parallel to the 12-inch side. These allow air to pass freely through the plant press for rapid drying of the pressed plant.

Blotters or Newsprint

Blotters or newsprint are in direct contact with the plant and absorb juices that may be squeezed from it as it is pressed. If 12-by-18-inch blotters are not available, pages of newsprint folded in half are approximately 12-by-14 inches and are suitable substitutes. Three or four pages of newsprint folded in half will perform the same function as a blotter.

Mounting Paper

Some collectors may wish to mount their plant collections on the pages of scrapbooks. An advantage of this is the great variety in the types of scrapbooks available and the ease of displaying and viewing the collection. A major disadvantage is that most scrapbook pages are smaller than standard herbarium sheets, which are $11^1/_2$-by-$16^1/_2$ inches. This is the size of mounting paper used in all professional herbaria. The personal herbarium of the collector will be of greater value if its specimens are compatible with those of professional herbaria. Mounting paper can be purchased at biological supply houses.

Labels

The sheet on which the specimen is mounted must have a label. Commercial mounting paper can be purchased with the label already printed in the lower-right-hand corner of the sheet. Plain paper is less expensive and standardized printed labels can be purchased separately or easily made. The label must provide several items of essential information. Obviously the first item should be the name of the plant. The manual that is used to identify the plant will give the botanical and common names and the label should carry both. The botanical name should be listed first. Sometimes it will be the only name since some species have no common names. In professional herbaria, the initials of the botanist who named the species are included as part of the botanical name.

The label should also give information in as much detail as possible about the location of the collecting site and the habitat from which the plant was collected. Finally, the name of the collector, the specimen number, and the date the collection was made should be listed.

The specimen number deserves a special mention. Some collectors keep a lifetime list of the plants they have identified or collected and number them consecutively from number one onward. Others prefer to start their numbering anew each year and designate the year of collection as 01–1, 01–2 then 02–1, 02–2, and so forth. The specimen number, environmental data, and site location will be provided by the field notebook.

All of this information can be recorded on a label about the size of a 3-by-5-inch card. If you make your own, four labels can be typed on a sheet of $8^1/_2$-by-11-inch paper. The following is suggested as a model.

HERBARIUM OF JANE DOE

Botanical Name_____

Common Name _____

Family _____

Locality _____

Habitat_____

Collector _____

Date _____No. _____

Adhesive

The function of the adhesive is to attach the specimen to the mounting paper. White glue such as Elmer's glue is probably the most convenient for the individual collector. It is readily available from many stores, is very effective, and is used by many professional botanists. Some collectors prefer thin strips of an adhesive linen tape to attach the specimen. This type of adhesive is available at most office supply stores. Transparent plastic tape is unsatisfactory because it dries and yellows with age.

Pressing the Specimen

Some collectors carry a plant press into the field and press the specimens as soon as they are collected. Others prefer to transport the specimens to home base where conveniences such as work tables may be available. Regardless of the location, there is a recommended routine for the process as described below.

1. The bottom frame of the press should be placed on the ground or on a table.

2. A corrugated cardboard ventilator is the placed on the frame.

3. This is followed by a blotter, or in the absence of blotters, several pages of newsprint folded in half.

4. The plant specimen to be pressed is placed on one half of a page of folded newsprint. It should be spread carefully so that flowers are unobstructed and there is a minimum overlapping of leaves. One or two leaves should be turned over with the bottom side up, since features of the leaf undersides are sometimes important for identification. If a specimen is too large to fit easily on one half of a page of newsprint, it can be bent to form a V, or if still larger, bent again to form an N. Then the other half of newsprint page is folded over the specimen. The name of the plant or its number is written on the outside of the folded newsprint.

5. A blotter, or two or three pages of folded newsprint are then placed on top of the newsprint containing the plant.

6. Another corrugated cardboard ventilator is placed on the blotter or newsprint.

7. The process can now be repeated for other specimens in the same

order: ventilator, blotter, specimen, blotter, ventilator. The plant press can be used to dry more than one plant at a time.

8. When all specimens have been so prepared, the top frame of the press is placed on the stack and the straps tightened around each end. Apply as much pressure as possible in tightening the straps. Having someone stand on the press while tightening is helpful.

Drying

The faster the specimen dries, the less likelihood that there will be discoloration of the flowers and leaves. Pressing a plant in a book is not recommended because the specimen dries slowly with practically no air circulation, usually resulting in discoloration of not only the plant but the pages of the book as well. In a plant press, depending on the temperature, humidity, and the size of the specimen, it will dry in five to ten days. After the first twenty-four hours, it can be examined to rearrange flowers or to smooth wrinkles.

If faster drying is desired, the plant press can be positioned over a mild source of heat so that warm air rises through the channels provided by the corrugations of the ventilators. The most convenient source of heat is probably an ordinary light bulb. The press can be placed between two chairs with the bulb at least one foot below the corrugations. Only mild heat is recommended because overheating may cause the specimens to turn brown. The plants will dry in two or three days with this arrangement.

If blotters and ventilators are not available, the plant press can still be useful. On the bottom frame, place a stack or three or four pages of folded newsprint. On top of these, place the folded page holding the plant to be pressed. Add another stack of newsprint similar to the first. At this point other specimens can be added following the same procedure. The top frame of the press can now be positioned and the straps tightened. It may take a little longer for the plants to dry by this method, but the porosity of the newsprint will provide enough aeration to prevent discoloration. To hasten the drying process, the newsprint can be changed after the first twenty-four hours.

Mounting

When the plant is removed from the press, it is ready to be attached, or mounted, on a sheet of paper. The traditional method of mounting is to coat a pane of glass with a thin layer of brown glue, lay the dried specimen on the pane to pick up glue, then place it on the mounting paper. This is a satisfactory method for a large professional herbarium with many plants to mount, but it is not suitable for the individual collector who may wish to mount only one or two plants at a time.

Using white glue that can be squeezed through a nozzle from a tube or other container, the individual collector can apply dots of glue to several places on the underside of the specimen. After positioning it on the paper, dots of glue can be applied to other points as needed. Sometimes a thin string of glue, when it dries across a leaf or other delicate part, will effectively pin it to the paper. An advantage of using glue is that usually it will last as long as the paper or the plant. A disadvantage is that the plant can never be removed from the sheet.

An alternate method of mounting is to use linen gummed tape. Thin strips of tape can be placed across stems and leaves at critical points to hold the plant on the paper. This method has the advantage of allowing the removal of the specimen from the paper at some future date. A disadvantage is that over a long period of time the tape may dry and lose its adhesiveness.

Attaching the Label

Attaching a label, complete with the name or identifying number of the plant, must be a part of the mounting routine. The most convenient ones are those that are already printed on some grade of commercial herbarium paper. However, gummed labels can be purchased. If you make your own, they can be attached with the same glue that was used to attach the specimen.

Protecting and Storing the Collection

There are two major threats to any herbarium: fungi and insects. Preventing contact of the specimen with moisture is the key to controlling fungal growth. If the mounts are dry at all times and stored in an area that has a consistently low humidity, the threat of fungal attack is greatly reduced. A greater problem is often caused by insects. Even if the plants are completely dry, an infestation may occur. Among the most damaging of the insect pests are several species collectively called dermestid beetles. They are very small beetles that in the larval stages feed on dry plant tissue. The collection should be inspected at least three times a year for indications of fungal or insect damage.

Professional herbaria store their collections in air-tight metal cabinets. These can be purchased from biological supply houses, but they are very expensive and probably impractical for the individual collector. Professional herbaria also use large manila folders, called species covers, to hold all the specimens of each species. While these are convenient, they are not essential, and, in their stead, the collector can use folded newsprint pages. Any appropriately sized cabinet or even cardboard boxes will serve as storage facilities. They can be made approximately airtight by splitting large plastic trash bags and tacking or gluing them in as liners. If the collection is mounted in scrapbooks, these can be stored in large plastic bags. It is worth repeating that whatever the storage facility, the storage area should be well ventilated and dry.

There are several types of fumigants that can be used to protect the herbarium from insects. The easiest to acquire is probably paradichlorobenzene, or PDB, which can commonly be purchased as either moth crystals or moth balls. If the storage cabinets or boxes have reasonably tight closure, a small cloth bag of crystals or perforated bags of moth balls can be placed in each compartment. Like scrapbooks, smaller collections mounted on individual sheets can be stored in large plastic bags into which crystals or moth balls have been inserted. The chemicals should be renewed about every four months. If the herbarium is stored at home, it should be placed in an area where family members will not be constantly exposed to PDB fumes.

Some botanists have suggested an alternate method of protecting the plant collection from infestation. Placing the mounts into a freezer for twelve to fourteen days seems to be enough to kill insect pests. This has great appeal to many people because it eliminates the use of chemicals. A disadvantage may be that it requires the periodic availability of a considerable amount of freezer space.

Displaying the Collection

There are several ways that plant mounts may be prepared for display. Collectors often give presentations for school groups, Scout groups, 4-H clubs, or other organizations for young people. It may be desirable in these presentations to have specimens that can be handled by the audience. Young, eager, and curious hands can do a lot of damage to a dry and brittle mounted plant. For collections mounted in scrapbooks, the best books are those with individual pages that are removable and have plastic covers. These provide a measure of protection for the plant and are excellent for viewing.

Collectors who do not use scrapbooks may wish to laminate with plastic the mounts of the specimens they will use for a presentation. However, the cost of this option may be prohibitive. Perhaps a more realistic plan is to attach the plant mount to a standard corrugated cardboard ventilator, or other stiff cardboard to keep the specimen from bending, then wrap it tightly with adhesive plastic kitchen wrap. Plants prepared in this way are suitable for hands-on presentations to groups of all ages.

Sometimes special mounting is appropriate for specimens that are bulky or unusually attractive. For mounts that are flat, the whole sheet can be enclosed in a frame called a botanical mount. It consists of a stiff cardboard back with a glass front held together usually by black tape around the edges. The botanical mount may contain a thin layer of cotton to hold the mount in place. For bulky specimens such as those with thick stems, pine cones, or hard fruits, a type of frame known as a Riker mount is available. These are shallow cotton-filled boxes with a pane of glass on one side. The specimen is usually not mounted on a sheet of paper but is embedded and held in place by the cotton. Both of these types of mounts are expensive, but plants mounted in these ways are often so attractive that they can be displayed as wall hangings.

Special Plant Groups

Some collectors may wish to include examples of all major plant groups in their collections. The discussion in the preceding pages has been concerned mainly with methods of collecting and preserving seed plants. These methods are valid for most plants but some groups require a different type of treatment. The life histories and growth habits of the plant groups listed below are described in chapter 2.

Algae

Freshwater algae. Algae may be preserved as dry mounts or in preservative liquids. To make a dry mount, hold a piece of mounting paper in the water beneath the alga. Raise the paper very slowly, tilting it to let the water escape but not the alga. The paper containing the specimen can then be allowed to air dry or it can be put into a plant press. If a plant press is used, feel the alga and if it is sticky to the touch, cover it with a sheet of waxed paper to keep it from sticking to the newsprint. As it dries, the alga will become attached to the mounting paper, so an adhesive is unnecessary. When the specimen is dry, the paper can be trimmed and attached to a scrapbook page or other mounting sheet and a label added.

The method used to get a specimen of an alga on mounting paper can be used for other kinds of aquatic plants, many of which have finely dissected or have very thin leaves. When these are removed from the water, they collapse and become difficult to work with. A piece of mounting paper can be inserted beneath the plant underwater near the surface. The leaves can be arranged as desired and the plant lifted slowly out of the water. After allowing the mounting paper containing the plant to drain on newsprint, it should be placed in a plant press to dry. Drying may take longer than for nonaquatic plants, but the process can be hastened by changing the blotters or newsprint after twenty-four hours. When the aquatic plant is dry, it may be impossible to remove it from the paper on which it was collected. If so, follow the same procedure for preserving that was suggested for algae.

Small commercially available collecting bottles or home food containers such as baby food jars can be used to save plant specimens in liquid preservatives. A number of preservatives are available from supply houses. One that is especially good for algae is F.A.A., which is a combination of formalin, alcohol, and acetic acid. This is toxic, so it should be handled with care and kept out of the reach of children. A satisfactory preservative that is available from drug stores and supermarkets is rubbing alcohol. Although it is not harmful to the skin, it can be highly toxic if ingested. A problem with using liquid preservatives is the almost unavoidable loss by evaporation.

The bottles should be checked monthly and additional preservative added to keep a constant fluid level. The label can be glued to the outside of the container or included within.

Fungi

Slime molds. The best time to collect this group is when they are in the fruiting or spore-producing stage. The type of sporangium of a species is an important feature for identification. A small section of the rotting wood or soil on which the sporangia are growing should be collected. Since the fragile sporangia may be dry and brittle, care should be exercised in taking the sample. It can be stored in a small box with some moth crystals and with a label on the outside. Boxes of various sizes are available from suppliers, but small boxes around the home, such as those for jewelry, are just as good.

Mushrooms and other soft fungi. As with slime molds, the structures that are collected are fruiting or spore-bearing stages. They can be preserved by drying or in a liquid preservative. If stored in liquid, they should be cut near the ground and placed in the preservative immediately. In the absence of a commercial preservative, rubbing alcohol is satisfactory for this purpose.

If the soft fungi are to be preserved by drying, they should be dried as quickly as possible because they decay rapidly when moist. Applying some form of mild artificial heat will hasten the process and help prevent the onset of decay. After drying is complete, the specimens can be stored in labeled boxes of the appropriate size. Moth crystals should be included as protection from insect attacks. To avoid decay, it is especially important to keep these specimens dry.

Shelf fungi. Some of the shelf fungi are hard and woody. The collector needs only to break them from the log or stump on which they are growing. They may grow to a very large size, but the collector can choose the size that is best suited for the collection. When they are thoroughly air dried, they can be stored in boxes or plastic bags with PDB crystals.

Lichens. A good plan for collecting lichens is to use a plastic bag in the field, then transfer them to boxes or envelopes for storage. The specimen should include the small cup-like spore-producing structures of the fungal portion of the lichen. This is very important for identification. Care should be taken in their transport and storage because they are usually dry and brittle and easy to shatter. With fungi, as with all plant groups, detailed field notes should be made so that a complete label can be attached to each specimen.

Mosses and Liverworts

These are so small that any sample will contain several plants. As with lichens, they can be collected in the field with plastic bags. Complete specimens should have the spore-producing structures. In mosses this includes the green bottom portion with the stalk and capsule at the top. After allowing the specimens to air dry for two or three days, they can be stored with PDB crystals in appropriately labeled boxes or envelopes.

Ferns and Fern Allies

Ferns. Ferns, clubmosses, and horsetails can be pressed and mounted in the same way as seed plants. Some special notes on collecting will be useful. In all three groups, it is important that the specimen have spore-bearing structures. In many fern species, these are on the underside of the frond. In those that have dissected leaves, they are on the undersides of the leaflets. When pressing the fern leaf, be sure to turn a few leaflets over so the fruit dots, or sori, can be seen when the leaf is mounted. In other fern species, the spore-bearing structures are on separate stalks. These must be included for complete specimens.

Clubmosses. The clubmosses may have spore-bearing structures, or sporangia, in the axils of upper leaves or in cones at the tips of branches. The upright branches of some species arise from a horizontal stem that runs along the surface of the ground or just beneath the surface. The characteristics of this stem are sometimes important for identification, so a small section should be included with the specimen.

Horsetails. The main factor to keep in mind when collecting horsetails is that some species have a spore-bearing, or fertile, stem that appears early, then may disappear before the green vegetative shoot is fully developed.

Leaves, Leaf Skeletons, and Leaf Prints

Leaves. A collection of tree leaves is easy to make, and it can be especially helpful in hands-on presentations to youth groups. Fully developed leaves with no insect damage should be selected and pressed for mounting in the manner described for seed plants. They can be mounted one or more leaves per page, as the collector wishes. Leaf collections of woody plants are more useful because these plants can more reliably be identified by leaf characteristics than can herbaceous plants.

An alternate way to make a collection of leaves is to coat them with wax. After they are thoroughly dry, they can be pressed between layers of wax paper with a warm iron. This will apply a thin layer of wax to each side of the leaf. Prepared this way, the leaves can be either mounted on paper or stored in a box or plastic bag and they will last for years. Leaves that are at the peak of autumnal coloration can be waxed in this manner, and, although the color may fade after a few weeks, they can be used for attractive seasonal decorations.

Leaf skeletons. Another way of displaying a leaf collection is by leaf skeletons. Leaves are made up of a network of tiny veins. When the interconnecting tissue is removed, a beautiful lacy outline of the leaf remains. There is no quick or easy way to make a leaf skeleton, but when successfully done, the result justifies the effort.

There are chemical methods that can be used for skeletonizing leaves, but they require chemicals that may not be readily available to the individual collector. The easiest method is to let nature take its course. In a gallon of water, add two tablespoons of forest soil humus or other rich topsoil, to insure the presence of decay bacteria, and two tablespoons of sugar or heavy syrup to stimulate bacterial growth. Immerse leaves in the solution and let it stand for a month in a warm place. Then remove a leaf and wash it with a gentle stream of water to remove the soft tissue. If a month is not enough time for complete leaf decay, let the solution stand for another two or three weeks. The leaf skeletons can be mounted between glass plates for projection on a screen, or they can be attractively displayed in a scrapbook or on mounting sheets.

Leaf prints. A different way of making a leaf collection is to make a collection of leaf prints. This involves the use of printer's ink that can be acquired at print shops or office supply stores that sell ink for rubber stamp pads. Several methods have been recommended but one that is simple and produces good results is described below.

With a small soft brush, such as a pastry brush, apply printer's ink to a piece of masonite or other hard, smooth surface. Place the leaf, underside down, on the inked surface. Cover it with a sheet of newsprint and roll it very gently with only the weight of a rolling pin or a large drinking glass. The objective is to bring all the veins on the underside of the leaf into contact with the ink. Remove the leaf from the inked surface and carefully place the inked side on a sheet of white paper. Cover it with newsprint and again roll it gently making sure it does not move. The quality of the print will be determined by the amount of ink picked up by the underside of the leaf. With practice, this can be controlled by the amount of pressure applied when the leaf is on the inked surface. Prints with varying degrees of density can be made according to the wishes of the collector.

Photographing Plants

A camera can be a great asset for the plant collector. Almost any type of camera will suffice, but one that allows close-up focusing is recommended. It is also useful sometimes to have a camera with film that can be processed into slides for projection on a screen. Pictures of the collecting sites can also be taken. These provide an added measure of authenticity when attached to the mounted specimens. In addition, a picture of a plant as it grows in the wild is often helpful in identification.

Some individuals confine their collecting to what can be captured on film. This usually results in large numbers of color slides of wildflowers. As these accumulate, a system of organization for storage becomes a necessity. They may be organized by habitat, such as plants of wetlands or forests; or by geography, such as plants of New York or West Virginia. As familiarity with botanical classification increases, the collector may want to organize the slide collection by plant family, such as plants of the lily family or aster family. A collection of plant slides, organized in any way the collector chooses, is rewarding for private or public showings. Specially constructed boxes for slide storage can be purchased from supply houses.

The camera can also be used for time-lapse photography that will sometimes yield spectacular results. Taking daily photographs of a germinating seed or hourly photographs of a flower as it opens are exciting activities. Setting a camera tripod in exactly the same location for color photographs of collecting sites in each ecological season will provide valuable life-history information.

Scientific Supply Houses

A few supply houses are listed below where the equipment described in this chapter may be purchased.

Carolina Biological Supply Company
2700 York Road
Burlington, NC 27215

Central Scientific Company
3300 Cenco Parkway
Franklin Park, IL 60131

Frey Scientific
905 Hickory Lane
P.O. Box 8101
Mansfield, OH 44901

Wards
P.O. Box 92912
Rochester, NY 14692

Sargent-Welch Scientific
911 Commerce Court
Buffalo Grove, IL 60089

9

Activities and Investigations

1. Does Moss Grow on the North Side of a Tree Trunk?

In fiction, an individual lost in the forest frequently finds the way out by observing that moss grows on the north sides of tree trunks. Is this the truth or myth? Three types of growth may be seen on tree trunks: (1) green algae will appear as a green layer that does not rise above the surface of the bark, (2) mosses, small green plants with tiny stem-like and leaf-like structures, and (3) lichens. Crustose lichens occur in roughly circular patterns, are grey-green and, like green algae, do not rise above the surface of the bark. Foliose lichens, like mosses, stand above the surface of the bark (see chapter 2). After the following investigation, decide for yourself whether or not these organisms, or any combination of them, grow on the north sides of tree trunks.

Using a pane of glass or a piece of transparent plastic, outline a square of at least fifteen centimeters (cm) on a side and mark off a grid of square centimeters. A square with 15 centimeters on a side will have and area of 225 square centimeters (15 x 15 = 225). With a compass locate the north, south, east, and west sides of a tree trunk. For ease of measurement, the tree should be at least thirty centimeters (1 ft.) in diameter at breast height. Place the grid in the center of the trunk on each of the four sides at thirty centimeters above the ground. Estimate the total number of square centimeters covered by lichens, algae, and moss. For example, if there is growth covering fifty square centimeters on one side of the trunk, this can be expressed as a percentage (50÷225 x 100 = 22.2 percent). Repeat this process at 60, 90, and 120 centimeters (2, 3, and 4 feet) above the ground. Perform these measurements on several trees. The four sides can then be compared for the greatest percentage of plant cover. Now if you were lost in the woods would the growth on tree trunks help you find your way?

As a refinement to this investigation, place the grid on the northeast, northwest, southeast, and southwest sides of the tree trunks and determine the percentage of coverage at the same levels as before. Do these results change your conclusions? You can also determine the relative abundance of lichens, algae, and mosses. Is one more common than the others on any side of the tree trunks? Is the growth about the same at each of the levels above the ground?

2. Dormancy

In the northern portions of the temperate zone, many species of plants become dormant during the winter months. For these species, dormancy is more than simply a cessation of growth because of low temperatures. It is a chemically induced state that requires a period of chilling before continuation of growth. This has a profound influence on the plant's geographic distribution. If a plant with a chilling requirement is not exposed to the required minimum cold period, it may not break dormancy in spring; or if it does break dormancy, growth may be weak and the plant devoid of flowers. Thus a species may be limited in its southward distribution by length and severity of winter temperatures. One of the reasons that apple trees do not flourish in Florida is no doubt because winters there are not cold enough or long enough to fulfill the dormancy requirements of apple trees.

A. Seed Dormancy

The seeds of many herbaceous plants in the temperate zone are in a dormant state when mature and will germinate only after chilling. Dormancy is a complex subject and there is much about it that plant scientists do not yet understand. The minimum temperature required and the length of the period of chilling are not known for most species. In seeds of some species, a factor other than, or in addition to chilling may be necessary for breaking dormancy. The following simple investigation may yield information on dormancy in seeds.

In a wetland area to which you have easy access, locate a colony of wild plants that are common to your region. Preferably these should be species that produce an abundance of seeds. In September, before cool nights have begun, collect twenty to fifty seeds from one species and arrange them in conditions favorable for germination as described below. Observe the seeds daily for ten days for signs of germination and record your results. If mold has not covered the seeds, continue observations for an additional four days. After two weeks, the seeds can be discarded. Repeat this process after collecting new seeds in October, November, December, January, and February. In theory, if the species you have selected requires a period of chilling, the greatest percentage of germination should be observed with the seeds collected in the month after the chilling requirement has been satisfied.

An alternate approach to this investigation is to collect all the seeds in September and separate them into several groups, then expose each group to a different chilling period in a refrigerator. Other variations of this project can be designed and explored by the investigator.

When a seed germinates, the first visible sign is usually the emergence of the embryonic root. At the time when it first becomes visible to the naked eye, the seed has germinated. To create conditions favorable for germination, place the seeds on several layers of moist paper towels and cover them with a similar number of moistened towels. The towels and seeds can be placed inside a container such as a paper box with a layer of plastic on the bottom and a loose-fitting lid. The box should be stored in a location where the temperature does not drop below 16°C (60°F). Inspect the seeds every day and keep the paper towels moistened. When the first sign of germination appears, the seed has broken dormancy.

B. Bud Dormancy

The marker of ecological spring in the swamp forest is the swelling of buds (see chapter 7). This indicates they have broken dormancy. The event is preceded by increasing temperatures and day lengths. The required period of chilling may have been satisfied weeks or months earlier. The durations of chilling and warming are not known for most species of forest trees. Those that grow in northern regions may require longer periods than the same species in more southerly regions. It can be assumed that after the chilling requirement has been achieved, the buds will resume growth if temperature and light conditions are favorable.

The beginning of ecological winter is indicated by the fall of leaves. By the time this occurs the trees are in a state of dormancy. They will remain dormant at least until the required period of chilling has been achieved. The length of the period can be investigated in a superficial way by using twigs. When twigs are cut from the ends of branches and brought inside, they should be cut again underwater so that air bubbles do not plug the ends of water-conducting cells. The cut end should be kept in water, at room temperature, under twelve to fourteen hours of light per day. The twig should be inspected daily for at least a month. The first sight of any structure emerging from the bud signals the breaking of dormancy and the resumption of growth. If there is no change in a month, it is probably an indication that the chilling requirement has not been satisfied.

The first twig should be collected immediately after the fall of leaves. In some regions, leaves are on the ground by November 1. If this is the date of the first collection, others should be made, following the same procedure, on December 1, January 1, February 1, March 1, and April 1. If the buds resume growth on twigs collected on one of these dates, the chilling requirement has been satisfied by that date for that species.

3. Investigating Plants Using Pollen, Spores, and Gametophytes

A. Observing Moss Spore Dispersal and Germination

When moss spores germinate, they form green algal-like filaments called protonemas. Individual filaments are very difficult to see, but they may occur in networks made up of many strands. A diligent observer may be able to see them as green thread-like lines on moist soil usually near moss plants (see chapter 2).

The mechanism for the dispersal of moss spores may be seen with a simple hand lens. An intact moss plant complete with gametophyte and sporophyte can be collected and taken indoors. The capsule of the sporophyte has a cap that covers a ring of enfolded flap-like teeth. If the moss plant is placed under an incandescent lamp, the capsule will dry, the cap will pop off, and the teeth will flip outward expelling the spores. When the capsule is remoistened, the teeth will return to their enfolded position.

B. Cultivating Fern Gametophytes

Fern gametophytes commonly grow in areas where fern plants (fern sporophytes) are abundant, but they are difficult to find because they are so tiny (see chapter 2). They are fairly easy to cultivate.

Thoroughly clean a three—or four-inch clay flower pot and boil it for ten minutes or run it through a dishwasher, then pack it tightly with moist peat moss or shredded paper. Invert it in a dish of water so that the water is in contact with the contents of the pot. Add water to the dish as needed. Cover the dish of water and its inverted flower pot with a transparent glass or plastic bowl to keep out dust and fungal spores.

Fern spores are easily collected from sensitive and cinnamon ferns, both of which have separate specialized spore-bearing leaves. Sensitive fern is very common in damp areas. By gently tapping the spore-bearing leaves on a white sheet of paper, the brown spores can be collected. Uncover the inverted flower pot and sprinkle some of the spores on its surface. A common error is to dust the spores too heavily; try using a medicine dropper and spread them very sparsely.

Place the covered pot where it is out of direct sunlight, and mature fern gametophytes will develop in eight to ten weeks. Sporophytes may be visible in about eight weeks but will usually be plentiful in six months. The young fern plants can be transplanted in moist soil after their roots have developed.

C. Observing and Germinating Horsetail Spores

Horsetail spores are produced in cones at the ends of stems. Each spore has four long slender appendages, the elaters, which are wrapped around it as it develops. When the cone matures and the spores dry, the appendages extend like four twisted helicopter blades. This expansion ruptures the sporangium resulting in the spores' release. The expanded appendages serve as wings that aid in dispersal of the spores by air currents. A spore with its extended appendages can be seen at the lower range of vision with a hand lens of 15X or 20X magnification. When the spores are moistened, the elaters contract and wrap around the spore.

If the spores are sown on the surface of water in a bowl, they will germinate within a few days but will not develop to completion.

D. Cultivating Horsetail Gametophytes

A complete horsetail gametophyte can be grown if newly formed spores are sown on a layer of clean moist peat moss that has been boiled to kill fungal spores. The peat moss should be placed in a dish, sown lightly with spores, covered, and placed out of direct sunlight. If the peat moss is kept moist, gametophytes will develop in a few weeks.

The gametophyte looks like a tiny green pincushion about the size of a pinhead or ranging in size from one millimeter to one centimeter in diameter. The gametophyte produces both egg and sperm cells and several horsetail plants (sporophytes) may arise from the same gametophyte if more than one egg cell is fertilized.

E. Growing Horsetails From Cuttings

Cuttings from both field horsetail and scouring rush will root if planted in wet sand. The cuttings should be embedded in the sand to a depth that includes at least one joint of the stem. The stems should root in a week or so and will continue to grow if the sand is watered regularly.

F. Germinating Pollen

The pollen grain of seed plants is the immature male gametophyte, and it produces the sperm nucleus. In most seed plants, the sperm is nonmotile and is delivered to the vicinity of the egg cell by a pollen tube. In flowering plants, the pollen grain reaches the stigma by wind or an animal pollinator. It germinates there and grows through the style to the ovules in the ovary. After the egg cell in the ovule has been fertilized by the sperm nucleus carried in the pollen tube, the ovule becomes a seed. One pollen grain is required for the development of each seed produced. Many seed plants produce great quantities of pollen, especially those that are wind pollinated such as cattails.

In many species, the pollen grain will germinate and begin growing the pollen tube if sown on water. To observe this, fill a shallow bowl or pan with water and sprinkle some freshly collected pollen on the surface. In a day or two, some grains with beginning pollen tubes may be seen with a 14X or 20X magnifying hand lens.

G. Making Spore Prints of Mushrooms

Spore prints are easy to make and they can be useful in the identification of mushrooms. For instructions see
chapter 2.

4. Life History Investigations

As one becomes more interested in plants and plant growth, it is but a short step to life history investigations. These can be fascinating field activities that require keen observational skills and good record keeping. In addition, they can yield original information because life history studies have not been conducted on every species. In an investigation of this type, observations should be made on every aspect of the life history of the species. The following are suggestions for the events and characteristics to be recorded for flowering plants. This list is not intended to be all-inclusive. As you become familiar with the species, you may wish to add other observations.

A. Seeds

   1. Earliest date of germination
   2. Number of seed leaves (monocot or dicot)
   3. Type of fruit (fleshy or dry)
   4. Number of seeds per fruit
   5. Number of seeds per plant
   6. Size and weight of seeds
   7. Seed modifications for dispersal
   8. Date and method of seed dispersal
   9. Period of chilling needed before germination

B. Stems and Leaves

1. Rate of stem growth in centimeters per week
2. Number, location, and arrangement of branches
3. Date at which the stem stops growing in height
4. Height of stem at maturity
5. Date at which the stem achieves winter conditions (if woody)
6. Description of leaves (basal, stem, size, sessile, color, etc.)
7. Arrangement of leaves (alternate, opposite, whorled)
8. Type of leaves (simple, compound, lobed, entire, pinnate, palmate)
9. Insects that feed on the plant
10. Date of leaf fall or behavior as winter approaches
11. The manner in which the plant survives the winter
12. Nature of aboveground parts of the plant in winter

C. Flower

1. Date of appearance of first flower
2. Date of maximum blooming
3. Number and distribution of flowers
4. Number of flower parts
5. Agents of pollination
6. Date of pollen dispersal
7. Life span for each flower

D. Other Observations

1. Life span of plant (annual, biennial, perennial)
2. Characteristics of root system
3. Habitat of plant (woods, open fields, wetlands, etc.)
4. Type of vegetative reproduction, if any
5. Outstanding features of the plant
6. Stages of life cycle when it may be edible, medicinal, poisonous, etc.

5. Making Leaf Collections

Many plants, especially woody plants, can be identified by leaf characteristics alone. Leaf collections can thus serve as aids for identification. Waxed leaf collections can also be used for decorations. Alternate methods of making leaf collections are leaf prints and leaf skeletons. Leaf prints give an outline of the main veins of the leaf, and leaf skeletons show the entire lacy network of the internal vascular system. See chapter 8 for details on preparing leaf, leaf print, and leaf skeleton collections.

6. Study of a Local Wetland

The documentation of information about a local wetland can be a rewarding and productive activity. For large wetlands, such as those described in chapter 1, there is a considerable amount of information available. But for most smaller ones there is none. Data provided by this type of study can be used in many ways. In addition to the satisfaction it brings to the naturalist who compiled it, this information can be used to publicize the wetland and its importance in newspaper articles and Sunday supplements. It can also be used in presentations to inform and educate civic and school groups.

The following outline is by no means exhaustive and the investigator may have other items to add to the study.

1. Name and geographic location of the wetland.

2. Size of area in acres or hectares. In the United States, USGS maps will be of assistance in determining size.

3. Ownership of the property. It may be privately owned or by federal, state, or municipal governments, or by a conservation organization such as the Nature Conservancy. Permission to do the study must be acquired from the owner.

4. Determine the nature of the wetland. Is it a swamp, marsh, bog, fen, swamp forest, or some combination of these?

5. Determine the origin of the wetland. Is it natural or man-made? What is the source of water? Is it fed and drained by streams?

6. Compile a list of plants with common and botanical name to include the following groups.

A. Ferns
B. Clubmosses

C. Horsetails

D. Seed Plants

    (1) woody plants

        (a) shrubs

        (b) trees

    (2) herbaceous plants

        (a) aquatic plants

           submergents

           floating plants

           emergents

        (b) other damp soil herbaceous plants

E. Mosses and Liverworts

F. Algae and Fungi

7. Compile a list of animals (vertebrates)

A. Mammals

B. Amphibians

C. Reptiles

D. Birds

8. Physical status

A. What changes are taking place in this wetland? Is it shrinking or expanding or neither?

B. What are the threats to its ecology?

C. Is it protected by federal or state laws? If so, are these being violated?

9. Values of this wetland

A. Recreational

    (1) canoeing

    (2) photography

    (3) bird watching

B. Nature Study

C. Flood Control

D. Water Purification

7. Determining the Number of Sphagnum Plants per Square Meter on a Bog Surface

This exercise is of little practical value but it is an interesting activity with numbers. Place a grid similar to the one recommended for exercise 1 on the bog surface and count the number of sphagnum plants in each square centimeter. Count a plant only when more than half of it is within the square. Total the number of plants counted in each square and divide by 225 to get the average number per square centimeter. Multiply that number by 10,000 to get the number of plants in a square meter of bog surface.

8. Culturing the Water Mold Saprolegnia

Saprolegnia is an aquatic fungus whose motile reproductive spores, called zoospores, are usually present in freshwater. It is able to become established on the external mucus of fish, and infections have been observed in sport fish such as salmon, trout, and perch. It is especially dangerous in wounds where it grows rapidly, and if it invades the blood vessels and spinal cord, death may result within twenty-four hours. Species of this fungus have caused great economic damage by attacking fish roe in commercial hatcheries. It may also infect amphibians and their eggs as well as insects and other aquatic organisms. These fungi are easy to cultivate because they grow rapidly on many kinds of organic matter. The following procedure should produce results within a relatively short period of time.

Collect a sample of clear pond water. Place any of the following in the water: a dead fly, a sesame seed, a boiled grain of wheat or corn. In two or three days, there should be a white fluffy growth of fungal mycelium around the object.

9. Using Leaf Characteristics to Distinguish Between Common Arrow-head, Arrow-arum, and Pickerel-weed

These species are common emergent aquatic plants that are widespread in eastern North America. They each have leaves that are similar in several ways: they have long petioles, they are arrow-head shaped, and they have two basal lobes. For these reasons, they are sometimes mistaken for one another. However, the venation in the leaves of each species differs greatly.

A. In common arrow-head (*Sagittaria latifolia*, fig. 9.1), all the veins in the leaf branch from the point of attachment of the petiole. The lower veins converge at the tips of the lobes, the upper ones converge at the tip of the leaf.

B. In arrow-arum (*Peltandra virginica*, fig. 9.1), each basal lobe has a heavy central vein.

C. In pickerel-weed (*Pontederia cordata*, fig. 9.1), all the veins arise from the point of attachment of the petiole and follow the contour of the leaf through each basal lobe, then converge at the tip of the leaf.

10. Field Identification of Grasses, Sedges, and Rushes

Members of these three families are very commonly found in wetlands throughout the world. They are large families, and identification to the species level in each is often based on technical traits that are daunting not only to beginners but to experts as well. Even identification to the family level can be challenging to the inexperienced because the plants in these families are all grass-like in general appearance. The following suggestions are offered as fairly simple ways to distinguish between them in the field. This method of identification is not scientifically foolproof, and it will not

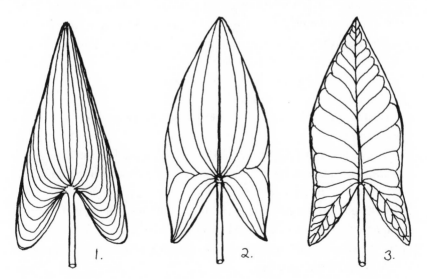

9.1. Leaves of (1) Pickerel-Weed (2) Common Arrow-Head and (3) Arrow-Arum

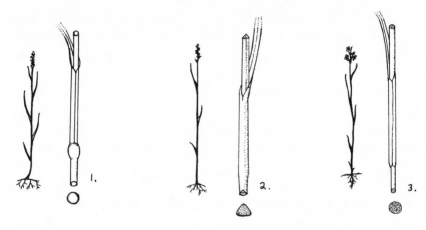

9.2. Leaves and stems of (1) Grasses, (2)Sedges, and (3) Rushes

work for every species of each family. If identification to the genus or species level is desired, the reader should consult the books listed in the bibliography by Gleason and Cronquist, Fasset, and Muenscher.

The grass family (*Poaceae*, fig. 9.2) is one of the largest of plant families with at least four thousand species worldwide. It is of great economic importance because it includes the cereal grasses such as wheat, oats, rice, and corn, which provide food for most of the world's population. Characteristics of the grass family that can be observed in the field are as follows.

A. The areas of stems where leaves are attached are called nodes. In grasses, the nodes are slightly enlarged.

B. Grass stems are usually circular in cross section.

C. Grass stem internodes are hollow.

D. The base of the grass leaf forms a sheath around the stem that is open on one side.

The sedge family (*Cyperaceae*, fig. 9.2) is less abundant than the grasses but still very common in wetlands, especially in cooler portions of both the southern and northern hemispheres. The tissue in the center of the stems is called the pith. All sedges have it, and the pith of one species, papyrus plant (*Cyperus papyrus*), was used by the Egyptians for making paper. Characteristics of the sedge family that can be recognized in the field are as follows.

A. Nodes are not enlarged.

B. The center of the stem is solid not hollow.

C. The stem is triangular in cross section.

D. The base of the leaf forms a sheath that is closed around the stem.

The rush family (*Juncaceae*, fig. 9.2) includes only two genera in North America, the rushes (*Juncus spp.*) and the wood rushes (*Luzula spp.*). The wood rushes are fewer in number and are found mostly in open spaces or woodlands. *Juncus* is the largest and most common genus and the one most likely to be seen in eastern wetlands. One of the differences between these genera is that the sheath around the stem is closed in the wood rushes and open in the rushes. Characteristics of the rush family, specifically the genus *Juncus*, that can be recognized in the field are as follows.

'A. Nodes are not enlarged.

B. The center of the stem is solid, not hollow.

C. The stem is circular in cross section.

D. The base of the leaf forms a sheath around the stem that is open on one side.

Glossary

adaptation: A characteristic of an organism that contributes to its survival under the conditions of the environment.

aeration: The process of adding air.

alternate leaf: A leaf arrangement in which there is one leaf at each node.

annual plant: A plant that completes its life cycle in one year and then dies.

anther: The part of the stamen that produces pollen.

axil: The angle between the leaf and the stem.

biennial plant: A plant that lives for only two years, producing flowers and seeds in the second year.

biomass: The total amount of organic matter produced by a plant or in a given area.

blade (of a leaf): The flat expanded portion of a leaf.

calyx: The sepals collectively.

canopy: The continuous cover over the forest floor formed by the crowns of the tallest trees.

climax vegetation: The final stages in ecological succession, composed of species that can reproduce themselves rather than being replaced by other species.

clone: A plant that is genetically identical to its parent plant.

compound leaf: A leaf in which the blade is divided into leaflets.

corolla: The petals of a flower.

cuticle: A waxy covering on all the aboveground parts of a plant.

deciduous plants: Plants that lose their leaves at the end of the growing season in contrast to evergreen plants.

diploid: Containing two full sets of chromosomes, one set from the egg and one from the sperm. Zygotes and sporophytes normally are diploid.

disk flower: A tiny flower on the central disk in the flower head of the aster family, as distinct from ray flowers.

dissected leaf: A leaf that is divided into many narrow segments as in some ferns.

ecological succession: The natural replacement of one plant community by another, culminating in climax vegetation.

ecosystem: A community of living things and all the physical factors that make up the environment.

fertile: Capable of sexual reproduction.

fertilization: The union of two haploid gametes resulting in a diploid zygote.

frond: The leaf of a fern.

gamete: A haploid sex cell such as an egg or a sperm.

gametophyte: A haploid gamete-producing structure or plant.

genus (plural genera): A group of closely related plants with a common ancestor. The first word of the two-word scientific name.

germinate: To resume growth, as a seed or a dormant spore or zygote.

girdle: To remove a ring of bark around the trunk of a tree.

ground water: The water in the ground in the saturated zone or below the water table.

habitat: The environment of an organism or a community.

haploid: Having only one set of chromosomes; for example, in gametes, spores, and gametophytes.

herb: A nonwoody plant that dies back to the ground at the end of the growing season; plants used in medicine or for seasoning.

herbaceous: Having the characteristics of a herb; green, leafy, with nonwoody tissue.

herbalist: One who collects, sells, or prescribes medicinal herbs.

hydrophyte: A plant that grows in a wet environment where it is partially or completely submerged.

internode: The portion of the stem where no leaves are attached; the space between nodes.

intertidal zone: That part of the coast between low tide and high tide.

leaflet: One of the divisions that make up a compound leaf.

mesophyte: A plant that grows in environmental conditions that are intermediate with regard to moisture; between hydrophytic and xerophytic.

micelle: A very tiny soil particle.

morbid: Unnatural; not sound or healthy; diseased.

nectar: A sweet fluid produced by flowers to attract pollinators.

node: The location on a stem where one or more leaves are attached.

opposite leaves: A leaf arrangement with two leaves per node; leaves attached in pairs.

organic matter: Living or once-living tissue; carbon compounds formed by living things.

ovary (plant): The enlarged basal portion of the pistil that contains the ovules and develops into the fruit.

ovule: An embryonic structure inside the ovary that will become a seed.

palmate: In compound leaves, an arrangement in which leaflets are attached at one point and radiate outward as the fingers from the palm of the hand.

perennial plant: A plant that lives for more than two years; not annual or biennial.

petals: The colorful segments of flowers that attract pollinators.

petiole: The stalk of a leaf.

phyte: A suffix that means plant; often preceded by a descriptive prefix such as hydrophyte, xerophyte, gametophyte.

pinnate: A form of compound leaves in which the leaflets are attached to each side of a central midrib.

pioneer species: The first plants to colonize bare soil or rock.

pistil: The female reproductive part of a flower; the seed-bearing part consisting of a style, stigma, and ovary.

plant community: All the plant species growing in an area.

pollination: The transfer of pollen from the anther to the stigma.

potherb: A herbaceous plant that is edible when cooked, including the leaves and sometimes the stem.

radial: Spreading outward from a central point.

ray flower: A marginal strap-shaped flower of the aster family.

rhizome: A creeping, horizontal, underground stem.

salinity: The degree of saltiness.

sepals: The outermost parts of the flower, usually green and leaf-like, that cover the outer parts of the bud.

shrub: A woody perennial not as large as a tree, usually with more than one stem.

simple leaf: A leaf that has a blade not divided into leaflets.

sp.: An abbreviation that follows the name of a genus and indicates a single unnamed or unknown species; *Acer sp.*

species: A group of organisms that can interbreed with one another but not with members of other species.

sporophyte: A diploid plant that produces haploid spores in plants that have alternation of generations.

spp.: An abbreviation that follows the name of a genus and indicates more than one unnamed or unknown species.

stamen: The male or pollen-producing structure of a flower consisting of an anther and a filament.

stigma: The part of the pistil that receives pollen and where the pollen germinates.

style: Usually a slender stalk that attaches the stigma to the ovary.

subspecies: A geographical race of a species.

substrate: Foundation material that makes up a given area of the earth. For example, a bog has an organic substrate.

succession: See ecological succession.

succulent: Thick, juicy, fleshy; for example, the leaves and stems of plants adapted for dry environments.

summergreen: A term sometimes used to describe the eastern deciduous forests that are green in the summer only in contrast to evergreen.

terrestrial: Living on dry land as opposed to in an aquatic environment.

thallus: A plant body that is not modified into root, stem, and leaf; for example, some of the liverworts.

transpiration: The loss of water by evaporation from the surface of plants.

turion: A bulb-like structure that serves as a winter bud.

understory trees: Trees that grow beneath the canopy of a forest but do not become part of the canopy.

vegetation: All the plants.

viable: Alive and capable of growth; for example, a seed.

water table: The top surface of the ground water.

whorled leaves: An arrangement of leaves with three or more attached at a node.

windfall: Trees blown down by the wind.

wort: A suffix that means plant.

xerophyte: A plant adapted to live under dry conditions.

zygote: A diploid cell formed by the union of two haploid gametes.

Bibliography and Further Reading

Abramaovitz, Janet N. 1996. *Imperiled Waters, Impoverished Future: The Decline of Freshwater Ecosystems.* Worldwatch Paper 128. Washington, D.C.: Worldwatch Institute.

Agriculture Research Service of the United States Department of Agriculture. 1970. *Common Weeds of the United States.* New York: Dover Publications.

Anderson, Frank J. 1997. *An Illustrated History of the Herbals.* New York: Columbia University Press.

Bailey, Liberty Hyde. 1933. *How Plants Get Their Names.* New York: Macmillan.

Barbour, M. G., J. H. Burk, and W. D. Pitts. 1980. *Terrestrial Plant Ecology.* Menlo Park, Calif.: Benjamin/Cummings Publishing.

Bell, C. Richie, and B. J. Taylor. 1982. *Florida Wild Flowers.* Chapel Hill, N.C.: Laurel Hill Press.

Benson, Lyman. 1979. *Plant Classification.* Lexington, Mass: D. C. Heath and Co.

Berlin, Brent. 1973. "Folk Systematics in Relation to Biological Classification and Nomenclature." *Annual Review of Ecology and Systematics*, vol. 4. Palo Alto, Calif.: Annual Reviews.

Billings, W. D. 1964. *Plants and the Ecosystem.* Belmont, Calif.: Wadsworth Publishing.

Bold, Harold C., C. J. Alexopoulos, and T. Delevoryas. 1987. *Morphology of Plants and Fungi.* New York: Harper and Row.

Braun, E. Lucy. 1950. *Deciduous Forest of Eastern North America.* New York: Macmillan.

Brayshaw, T. Christopher. 1996. *Plant Collecting for the Amateur.* Victoria, British Columbia: Royal British Columbia Museum.

Brown, Lester, et al. 1990. *State of the World 1990.* New York: W. W. Norton.

Buchholz, Rogene A. 1998. *Principles of Environmental Mangement.* 2d ed. Upper Saddle River, N.J.: Prentice Hall.

Campbell, F. T. 1980. "Conserving Our Wild Plant Heritage." *Environment* 22, no. 9: 14–20.

Carlson, Eric, D. Cusick, and C. Taylor. 1992. *The Complete Book of Nature Crafts.* Emmaus, Penn.: Rodale Press.

Charas, Daniel D. 1992. *Lessons From Nature.* Washington, D.C.: Island Press.

Christensen, N. L., R. B. Burchell, A. Liggett, and E. L. Simms. 1981. "The Structure and Development of Pocosin Vegetation." In *Pocosin Wetlands,* ed. C. J. Richardson. Stroudsburg, Penn.: Hutchinson Ross Publishing.

Cobb, B. 1977. *A Field Guide to the Ferns and Their Families of Northeastern and Central North America.* Boston, Mass.: Houghton Mifflin.

Coffey, Timothy. 1993. *The History and Folklore of North American Wildflowers.* New York: Facts on File.

Cook, C. D. K. 1990. "Origin, Autecology, and Spread of Some of the World's Troublesome Weeds." In *Aquatic Weeds: The Ecology and Management of Nuisance Aquatic Vegetation,* ed. A. H. Pieterse and K. T. Murphy, 31–38. New York: Oxford University Press.

Cope, Edward A. 1992. *Pinophyta (Gymnosperms) of New York State.* New York State Bulletin No. 483. Albany, N.Y.

Core, Earl L. 1955. "Cranberry Glades Natural Area." *Wildflower* 32: 65–81.

Corroll, Don. 2000. "Purple Loosetrife: Controlling a Colorful Nuisance." *New York State Conservationist* 31, no. 1: 12–14.

Courtenay, B., and H. H. Burdsall, Jr. 1984. *A Field Guide to Mushrooms and Their Relatives.* New York: Van Nostrand Reinhold.

Cox, Donald D. 1996. *Seaway Trail Wildguide to Natural History.* Sackets Harbor, N.Y.: Seaway Trail Foundation.

Cox, Donald D. 1995. *Common Flowering Plants of the Northeast.* Albany, N.Y.: State University of New York Press.

Crawley, M. J., ed. 1986. *Plant Ecology.* Boston, Mass.: Blackwell Scientific Publications.

Croom, Edward M. 1983. "Documenting and Evaluating Herbal Remedies." *Economic Botany* 37, no. 1: 13–27.

Crum, Howard A., and L. E. Anderson. 1981. *Mosses of Eastern North America.* New York: Columbia University Press.

Cutter, Susan L., H. L. Renwick, and W. H. Renwick. 1985. *Exploitation, Conservation, Preservation.* Totowa, N.J.: Rowman and Allanheld Publishers.

Daiber, Franklin C. 1986. *Conservation of Tidal Marshes.* New York: Van Nostrand Reinhold.

Darlington, H. Clayton. 1943. "Vegetation and Substrate of Cranberry Glades, West Virginia." *Botanical Gazette* 104, no. 3:371–93.

Daubenmire, R. F. 1959. *Plants and Environment.* 2d ed. New York: John Wiley and Sons.

Dressler, R. L. 1981. *The Orchids: Natural History and Classification.* Cambridge, Mass.: Harvard University Press.

Edens, David Lee. 1973. "The Ecology and Succcession of Cranberry Glades, West Virginia." Ph.D. thesis, North Carolina State University.

Ehrlich, Paul, and Anne Ehrlich. 1981. *Extinction: The Causes and Consequences of the Disappearance of Species.* New York: Random House.

Fahn, Abraham, and E. Werker. 1972. "Anatomical Mechanisms of Seed Dispersal." In *Seed Biology,* vol. 1. Ed. T. T. Kozlowski, 151–221. New York: Academic Press.

Fasset, N. C. 1976. *Manual of Aquatic Plants,* revised by E. C. Ogden. Madison: University of Wisconsin Press.

Fay, Peter. 1983. *The Blue Greens.* Baltimore, Md.: Edward Arnold.

Fernald, Merritt L. 1970. *Gray's Manual of Botany,* 8th ed. corrected printing. New York: D. Van Nostrand.

"Fire Effects Information System." Prescribed Fire and Fire Effects Research Unit, Rock Mountain Research Station. 1998 <http://www.fs.fed.us/database/fires>.

Gibbons, Euell. 1966. *Stalking the Heathful Herbs.* Brattleboro, Vt.: Alan C. Hood and Co.

Gibbons, Euell. 1962. *Stalking the Wild Asparagus.* New York: David McKay.

Gibbons, Euell, and G. Tucker. 1979. *Euell Gibbons Handbook of Edible Wild Plants.* Virginia Beach, Va.: Unilaw Library Press.

Given, David R. 1994. *Principles and Practice of Plant Conservation.* Portland, Ore.: Timber Press.

Glaser, P. H. 1987. *The Ecology of Patterned Boreal Peatlands of Northern Minnesota: A Community Profile.* Report 85 (7.14). Washington, D.C.: U.S. Fish and Wildlife Service.

Gleason, Henry A., and A. Cronquist. 1991. *Manual of Vascular Plants of Northeastern United States.* Bronx, N.Y.: New York Botanical Garden.

Gleason, H. A., and A. Cronquist. 1964. *The Geography of Plants.* New York: Columbia University Press.

Glob, P. V. 1969. *The Bog People: Iron-Age People Preserved.* Ithaca, N.Y.: Cornell University Press.

Good, E. G., D. R. Whigham, R. L. Simpson, and C. G. Jackson. 1978. *Freshwater Wetlands: Ecological Processes and Management Potential.* New York: Academic Press.

Hale, M. E. 1979. *How to Know Lichens.* 2d ed. Dubuque, Ia.: Wm. C. Brown.

Hammer, Donald A. 1992. *Creating Freshwater Wetlands.* Chelsea, Mich.: Lewis Publishers.

Hardin, James W., and J. M. Arena. 1974. *Human Poisoning from Native and Cultivated Plants.* Durham, N.C.: Duke University Press.

Harper J. L., P. H. Lovell, and K. G. Moore. 1970. "The Shapes and Sizes of Seeds." In *Annual Review of Ecology and Systematics,* vol. 1. Ed. R. F. Johnston, P. W. Frank, and C. D. Michener, 327–56. Palo Alto, Calif.: Annual Reviews.

Hitchcock, S. T. 1980. *Gather Ye Wild Things.* New York: Harper and Row.

Howe, Henry F., and J. Smallwood. 1982. "Ecology of Seed Dispersal." In *Annual Review of Ecology and Systematics,* vol. 13. Ed. R. F. Johnston, P. W. Frank, and C. D. Michener, 201–28. Palo Alto, Calif.: Annual Reviews.

Jacobs, Barry L. 1987. "How Hallucinogenic Drugs Work." *American Scientist* 75: 386–92.

Johnson, Charles W. 1985. *Bogs of the Northeast.* Hanover, N.H.: University Press of New England.

Joosten, Titia. 1988. *Flower Drying with a Microwave: Techniques and Projects.* New York: Sterling Publishing.

Kaufman, Peter B., T. F. Carson, P. Dayanandan, M. L. Evans, J. B. Fisher, C. Parks, and J. R. Wells. 1991. *Plants: Their Biology and Importance.* 2d ed. Philadelphia, Penn.: Harper and Row.

Keeney, Elizabeth B. 1992. *The Botanizers.* Chapel Hill: University of North Carolina Press.

Kent, Donald M. 1994. *Applied Wetlands Science and Technology.* Boca Raton, Fla.: Lewis Publishers.

Ketchledge, E. H. 1970. *Plant Collecting: A Guide to the Preparation of a Plant Collection.* Syracuse, N.Y.: State University of New York College of Environmental Science and Forestry.

Kinghorn, A. Douglas. 1979. *Toxic Plants.* New York: Columbia University Press.

Kingsbury, John M. 1964. *Poisonous Plants of the United States and Canada.* Englewood Cliffs, N.J.: Prentice Hall.

Kivinen, E., and P. Pkarinen. 1981. "Peatland Areas and the Proportion of Virgin Peatlands in Different Countries." In *Proceedings of the Sixth International Peat Congress,* Duluth, Minn.

Koopowitz, Harold, and Hilary Kaye. 1983. *Plant Extinction: A Global Crisis.* Washington, D.C.: Stone Wall Press.

Kowalchik, Claire, and W. H. Hylton, eds. 1987. *Rodale's Illustrated Encyclopedia of Herbs.* Emmaus, Penn.: Rodale Press.

Krajick, Kevin. 1997. "The Riddle of the Carolina Bays." *Smithsonian* 28, no. 6: 44–55.

Krochmal, Connie, and Arnold Krochmal. 1973. *A Guide to the Medicinal Plants of the United States.* New York: Quadrangle/The New York Times Book Co.

Laderman, A. D. 1989. *The Ecology of the Atlantic White Cedar Wetlands: A Community Profile.* Biol. Report 85 (7.21). Washington, D.C.: U.S. Fish and Wildlife Service.

Lampe, Kenneth F., and M. A. McCann. 1985. *AMA Handbook of Poisonous and Injurious Plants.* Chicago, Ill.: American Medical Association.

Lewis, W. H., and M. Elvin-Lewis. 1977. *Medical Botany: Plants Affecting Man's Health.* New York: John Wiley and Sons.

Lincoff, G. H. 1981. *The Audobon Society Field Guide to North America Mushrooms.* New York: Alfred A. Knopf.

Litovitz, Toby L., L. R. Clark, and R. A. Solway. 1993. *Annual Report of the American Association of Poison Control Centers.* Washington, D.C.

Little, Ebert L. 1980. *The Audobon Society Field Guide to North American Trees (Eastern Edition).* New York: Alfred A. Knopf.

Lyman, Francesca, I. Mintzer, K. Courrier, and J. Mackenzie. 1990. *The Greenhouse Trap.* Boston, Mass.: Beacon Press.

MacFarlane, R. B. 1985. *Collecting and Preserving Plants for Science and Pleasure.* New York: Arco Publishing.

Marchand, Peter J. 1987. *Life in the Cold: An Introduction to Winter Ecology.* Hanover, N.H.: University Press of New England.

Marchand, Peter J. 2000. *Autumn: A Season of Change.* Hanover, N.H.: University Press of New England.

Mauseth, James D. 1991. *Botany: An Introduction to Plant Biology.* Fort Worth, Tex.: Saunders College Publishing.

Meeuse, B. J. D. 1961. *The Story of Pollination.* New York: Ronald Press.

Miller, David F., and G. W. Blades. 1962. *Methods and Materials for Teaching the Biological Sciences.* 2d ed. New York: McGraw-Hill.

Miller, G. Tyler, Jr. 1992. *Living in the Environment.* Belmont, Calif.: Wadsworth Publishing.

Millspaugh, Charles F. 1974. *American Medical Plants.* New York: Dover Publications.

Mitchell, D. S., and B. Gopal. 1991. "Invasion of Tropical Freshwaters by Alien Aquatic Plants." In *Ecology of Biological Invasions in the Tropics,* ed. P. S. Ramakrishnan, 139–54. New Delhi, India: International Scientific Publications.

Mitsch, William J. and J. G. Grosselink. 1993. *Wetlands.* 2d ed. New York: Van Nostrand Reinhold.

Momatiuk, Yva, and John Eastcott. 1995. "Creatures from the Black Lagoon." *Nature Conservancy* 45, no. 5: 25–29.

Momatiuk, Yva, and John Eastcott. "Liquid Land." *Audobon* 97, no. 5.

Muencher, Walter C. 1936. *Weeds*. New York: Macmillan.

Muenscher, Walter Conrad. 1972. *Aquatic Plants of the United States*. Ithaca, N.Y.: Cornell University Press.

Munzer, Martha E., and P. F. Brandwein. 1960. *Teaching Science Through Conservation*. New York: McGraw-Hill.

Murphy, Terrence M., and W. F. Thompson. 1988. *Molecular Plant Development*. Englewood Cliffs, N.J.: Prentice Hall.

Niering, William A. 1991. *Wetlands of North America*. Charlottesville, Va.: Thomasson-Grant.

Niering, William A. 1985. *Wetlands*. New York: Alfred A. Knopf.

Niering, William, and N. Olmstead. 1979. *Audubon Society Field Guide to North American Wildflowers (Eastern Region)*. New York: Alfred A. Knopf.

Niering, William A., and R. S. Scott. 1980. "Vegetation Patterns and Processes in the New England Salt Marshes." *Bioscience* 30, no. 5: 301–307.

Oosting, H. J. 1954. "Ecological Process and Vegetation of the Maritime Strand in the Southeastern United States." *Botanical Review* 20, no. 4: 226–62.

Palmer, E. Laurence. 1975. *Fieldbook of Natural History*. 2d ed. Revised by H. Seymour Fowler. New York: McGraw-Hill.

Peterson, Lee. 1977. *A Field Guide to Edible Wild Plants*. Boston, Mass.: Houghton Mifflin.

Prescot, G. W. 1978. *How to Know the Freshwater Algae*. 3d ed. Dubuque, Ia.: Wm. C. Brown.

Pritchard, Hayden N., and T. B. Bradt. 1984. *Biology of Nonvascular Plants*. St. Louis, Mo.: Times Mirror/Mosby College Publishing.

Reddington, Charles B. 1994. *Plants in Wetlands*. Dubuque, Ia.: Kendall/Hunt Publishing.

Richardson, Curtis J. 1983. "Pocosins: Vanishing Wastelands or Valuable Wetlands?" *Bioscience* 33, no. 10: 626–33.

Richardson, Curtis J., ed. 1981. *Pocosin Wetlands: An Integrated Analysis of Coastal Plain Freshwater Bogs in North Carolina*. Stroudsburg, Penn.: Hutchinson Ross Publishing.

Russell, Franklin. 1973. *The Okefenokee Swamp*. New York: Time-Life Books.

Saunders, C. F. 1948. *Edible and Useful Wild Plants of the United States and Canada*. New York: Dover Publications.

Schery, Robert W. 1972. *Plants for Man*. 2d ed. Prentice-Hall.

Simpson, Beryl Brintnall, and M. Conner-Ogorzaly. 1986. *Economic Botany: Plants in Our World*. New York: McGraw-Hill.

Simpson, Bland. 1990. *The Great Dismal: A Carolinian's Swamp Memoir*. Chapel Hill: University of North Carolina Press.

Stebbins, G. Ledyard. 1971. "Adaptive Radiation of Reproductive Characteristics in Angiosperms II: Seeds and Seedlings." In *Annual Review of Ecology and Systematics*, vol. 2. Ed. R. F. Johnston, P. W. Frank, and C. D. Michener, 237–60. Palo Alto, Calif.: Annual Reviews.

Steward, K. K. 1990. "Aquatic Weed Problems and Management in the Eastern United States." In *Aquatic Weeds: The Ecology and Management of Nuisance Aquatic Vegetation*, ed. A. H. Pieterse, and K. J. Murphy, 391–405. New York: Oxford University Press.

Thomas, Bill. 1979. *The Swamp*. New York: W. W. Norton.

Thompson, Daniel Q., Ronald L. Stuckey, Edith B. Thompson. "Spread, Impact, and Control of Purple Loosestrife *(Lythrum salicaria)* in North American Wetlands." U.S. Fish and Wildlife Service Research Center Home Page. 1987. <http://www.npwrc.usgs.gov/resource/1991/loosstrf/loostrf.htm>.

Tiner, Ralph W., Jr. 1987. *A Field Guide to Coastal Wetland Plants of the Northeastern United States*. Amherst: University of Massachusetts Press.

Tiner, Ralph W., Jr. 1988. *Field Guide to Nontidal Wetland Identification*. Annapolis, Md.: Maryland Department of Natural Resources and U.S. Fish and Wildlife Service.

Tippo, Oswald, and W. L. Stern. 1977. *Humanistic Botany*. New York: W. W. Norton.

Turner, Nancy J., and A. F. Szczawinski. 1991. *Common Poisonous Plants and Mushrooms of North America*. Portland, Ore.: Timber Press.

Van der Pijl, L. 1972. *Principles of Dispersal in Higher Plants*. 2d ed. New York: Springer-Verlag.

Weller, M. W. 1987. *Freshwater Marshes*. 2d ed. Minneapolis, Minn.: University of Minnesota Press.

Whitehead, Donald R. 1971. "Developmental and Environmental History of the Dismal Swamp." *Ecological Monographs* 42, no. 3: 301–15.

Williams, T. 1994. "Invasion of the Aliens." *Audubon* 96, no. 5: 24–32.

Index